D0108875

Techniques for Desert
Reclamation

ENVIRONMENTAL MONOGRAPHS & SYMPOSIA

Convener and General Editor
NICHOLAS POLUNIN, D. Phil, DSc, CBE
The Foundation for Environmental Conservation, Geneva, Switzerland

Modernization of Agriculture in Developing Countries: Resources, Potentials, and Problems

ISAAC ARNON, Hebrew University of Jerusalem Agricultural Research Service, Bet Dagan, and Settlement Study Centre, Rehovot, Israel

Stress Effects on Natural Ecosystems

Edited by
GARY W. BARRETT, Institute of Environmental Sciences and Department of Zoology, Miami University, Oxford, Ohio, USA
&
RUTGER ROSENBERG, Fishery Board of Sweden, Institute of Marine Research, Lysekil, Sweden

Air Pollution and Plant Life

Edited by
MICHAEL TRESHOW, Department of Biology, University of Utah, Salt Lake City, Utah, USA

Impounded Rivers: Perspectives for Ecological Management

GEOFFREY E. PETTS, Department of Geography, University of Technology, Loughborough, England, UK

Ecosystem Theory and Application

Edited by
NICHOLAS POLUNIN, Environmental Conservation, Grand-Saconnex, Geneva, Switzerland

Techniques for Desert Reclamation

Edited by
ANDREW GOUDIE, School of Geography, University of Oxford, England, UK

Techniques for Desert Reclamation

Edited by
Andrew S. Goudie
School of Geography,
University of Oxford

JOHN WILEY & SONS
Chichester · New York · Brisbane · Toronto · Singapore

Other Wiley Editorial Offices

John Wiley & Sons, Inc., 605 Third Avenue,
New York, NY 10158-0012, USA

Jacaranda Wiley Ltd, G.P.O. Box 859, Brisbane,
Queensland 4001, Australia

John Wiley & Sons (Canada) Ltd, 22 Worcester Road,
Rexdale, Ontario M9W 1L1, Canada

John Wiley & Sons (SEA) Pte Ltd, 37 Jalan Pemimpin   05-04,
Block B, Union Industrial Building, Singapore 2057

Library of Congress Cataloguing-in-Publication Data:

Techniques for desert reclamation /   edited by Andrew Goudie.
        p.   cm. — (Environmental monographs & symposia)
    Includes bibliographical references.
    ISBN 0 471 92179 3
    1. Desert reclamation. I. Goudie, Andrew. II. Series:
Environmental monographs and symposia.
S613.T44   1990
333.73′6153—dc20
                                                    90–30302
                                                         CIP

British Library Cataloguing in Publication Data:

Techniques for desert reclamation.
    1. Arid. Regions. Environment planning
    I. Goudie, Andrew
    307.1209154

    ISBN 0 471 92179 3

Typeset by Acorn Bookwork, Salisbury, Wiltshire
Printed and bound in Great Britain by Biddles Ltd., Guildford, Surrey

Contents

List of Contributors

ADAMS, W. M. *Department of Geography, University of Cambridge, Downing Place, Cambridge, CB2 3EN, England*

BISWAS, ASIT K. *International Water Resources Association, 76 Woodstock Close, Oxford, OX2 8DD, England*

BURLEY, J. *Director, Oxford Forestry Institute, University of Oxford, South Parks Road, Oxford, OX1 3PS, England*

COE, M. *Department of Zoology, University of Oxford, South Parks Road, Oxford, OX1 3PS, England*

GOUDIE, A. S. *School of Geography, University of Oxford, Mansfield Road, Oxford, OX1 3TB, England*

HUGHES, F. M. R. *Department of Geography, University of Cambridge, Downing Place, Cambridge, CB2 3EN, England*

LAL, R. *Department of Agronomy, The Ohio State University, Room 202, 2021 Coffey Road, Columbus, OH 43210-1086, USA*

MIDDLETON, N. *School of Geography, University of Oxford, Mansfield Road, Oxford, OX1 3TB, England*

RHOADES, J. D. *US Salinity Laboratory, 4500 Glenwood Drive, Riverside, CA 92501, USA*

WATSON, A. *Environmental Geosciences, 23 Prospect Street, Holliston, MA 01746, USA*

Series Preface

For civilization to survive in anything like its present form, the world's human population will need continuingly to increase and widen its knowledge of the ever-changing environment. Moreover, this knowledge will need to be closely followed by concomitant action to safeguard the Biosphere and so maintain the framework and chief structures of our life-support system. The *increase* in knowledge and awareness must come through observation, research, and applicational testing, its *widening* through environmental education, and the necessary concerted *action* through duly organized application of the knowledge that has been thus acquired and disseminated. Meanwhile such knowledge is causing increasingly grave concern to the extent that many serious people are beginning to fear for the very future of our world.

The environmental movement has long been an undefined but widely effective vehicle for increasing appreciation of the vital reality and fundamental importance of Man's and Nature's environment. As part of this movement, the World Campaign For The Biosphere, 1982– , and its adopting World Council for the Biosphere (ECB)—International Society for Environmental Education (ISEE), help to focus attention on the vulnerability of the Biosphere—that 'peripheral envelope of Earth together with its surrounding atmosphere in which living things exist naturally'. At the same time the environmental movement should stress our utter dependence on the Biosphere's health—as it constitutes our sole life-support system—and inculcate the necessity to foster it in every possible way, as is emphasized chronically in the quarterly journal *Environmental Conservation*.

To help to encourage such ideals and guide appropriate actions which in many cases are imperatives for Man *and* Nature, as well as to distil and widen knowledge in component fields of scientific and allied environmental endeavour, we founded and are now fostering an open-ended series of *Evironmental Monographs and Symposia*. This emanated from an invitation by the international publishers John Wiley & Sons, and consists of authoritative volumes of two main kinds: monographs in the full sense of being detailed treatments of particular subjects by from one to three leading specialists, and symposia by more than three specialist authors covering a particular subject between them under the guidance and editorship of a suitable specialist or up to three specialists (whether such a volume results in part or wholly from an actual 'live' conference or symposium, or consists partly or entirely of 'contributed' papers conforming to an agreed plan).

There seems to be virtually no end to the possibilities for this series: we are constantly getting or being given new ideas, and now have very many to think about and, in chosen cases, to work on. At the same time we hope to complement the existing SCOPE Reports, emanating from what in a sense is the world's environmental 'summit'. In our present series we have already published two successful editions (1981 and 1987) of *Modernization of Agriculture in Developing Countries: Resources, Potentials, and Problems*, by Professor I. Arnon, *Stress Effects on Natural Ecosystems*, edited by Professor Gary W. Barrett and Dr Rutger Rosenberg, *Air Pollution and Plant Life*, edited by Professor Michael Treshow, *Impounded Rivers: Perspectives for Ecological Management*, by Professor Geoffrey E. Petts, and my own *Ecosystem Theory and Application*, while others are in various stages ranging from merely being contemplated to actually in press.

The work that is about to be introduced, by the Professor and Head of Geography in the University of Oxford, is a worthy successor to the above.

Whether or not we shall in time come to cover, in however general a manner, the entire vast realm of environmental scientific endeavour must remain to be seen, though this was the gist of the distinguished publishers' original invitation and poses a challenge that we can scarcely forget. Meanwhile we believe we have decided on a constructive compromise with these most timely, open-ended series, in which we look forward to effective participation by more and more of the world's leading environmentalists.

NICHOLAS POLUNIN
(Convener and General Editor of the Series)
Geneva, Switzerland

Preface

Desert degradation is one of the great environmental problems that faces the human race, and it has now become a matter of international concern. Much debate has surrounded the question of causation (see Chapter 1), but a no less key issue is the question of solutions.

The purpose of this volume is to review some of the essentially technological solutions that may be appropriate to improve the status and productivity of arid lands. We recognize, however, the need for a companion volume which might explore some of the rather intractable political, social, and economic issues that underlie the problem of land degradation.

ANDREW GOUDIE
Oxford

CHAPTER 1

Desert Degradation

A. S. GOUDIE

INTRODUCTION: THE DESICCATION CONCEPT

One of the biggest environmental issues of the last two decades has been the question of desertification. Maps have been produced of the world which purport to show that huge areas are at risk from varying degrees of land degradation (Figure 1.1). It is widely reported that desertification affects about 65 million hectares of once-productive agricultural land and threatens the livelihood of 850 million people. Desertification and salinity have been identified by the World Bank as two of the nine major environmental problems in the world today. The purpose of this volume is to review some of the ways in which degraded arid lands may be reclaimed, and also to look at ways in which areas at risk may be protected from degradation.

The belief that areas within and on the margins of deserts are degrading, though current and fashionable, is not new. Over a century ago the concept of progressive desiccation arose (Goudie, 1972). This was based on two tenets: that wet conditions were a feature of the glacial phases of the Pleistocene and that aridity had increased since the warming of the Pleistocene ice sheets in the Holocene. In Central Asia the concept gained great momentum in the early years of this century, particularly after the great explorations of the Tarim Basin, Lop Nor and Tibet. For example, in 1903 Andrew Carnegie financed an expedition to Turkestan and Persia on which Ellsworth Huntington was a member. In 1905–6 Huntington made a second visit to Central Asia in which he visited India and Tibet.

On these journeys, details of which are presented in various reports, Huntington noted the evidence for ruined cities and abandoned settlements, and recognized after the Pumpelly Expedition that lake terraces indicated a 'gradual desiccation of the country from early historical times down to the present'. After the Barrett Expedition, which took him to the famed Tarim

Techniques for Desert Reclamation
Edited by A. S. Goudie
© 1990 John Wiley & Sons Ltd.

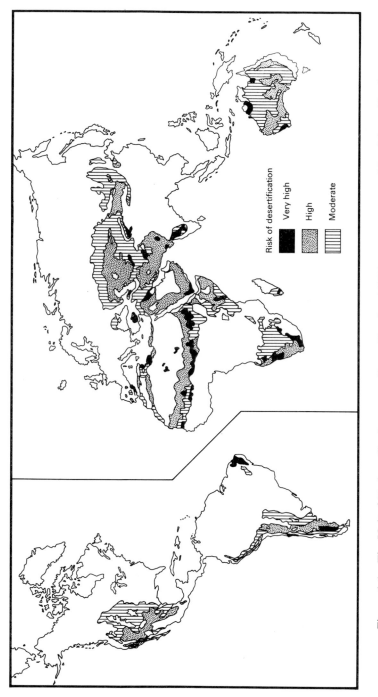

Figure 1.1   The United Nations Conference on Desertification (1977) map of areas at risk of desertification

Basin, he wrote his *Pulse of Asia* and moved away from any simple idea of progressive desiccation to a view embracing multiple large fluctuations. He believed that the pulsations of climate had served as a driving force in the history of Eurasia, forcing nomadic invaders to overrun their more civilized neighbours whenever the climatic cycle reached a trough of aridity. Such views were followed or accepted by numerous workers including Brooks (1922), Coching (1926) and Curry (1928). With regard to America itself the original simple progressive desiccation hypothesis was not particularly important in its impact and Bowman (1935), for example, who studied such evidence as lake fluctuations since the mid 1840s in the western USA was sceptical as to whether the country was in 'an ascending or descending phase of a long-range change in climate'.

At much the same time as Huntington put forward his views, Prince Peter Kropotkin, the anarchist–geologist–explorer–evolutionist, who travelled widely in the wastes of Central Asia with the Mounted Cossack Regiment, put forward a simple view of progressive desiccation, the like of which Huntington was to move away from (Kropotkin, 1904):

'Recent exploration in central Asia has yielded a considerable body of evidence, all tending to prove that the whole of that wide region is now, and has been since the beginning of the historic period, in a state of rapid desiccation. . . . It must have been the rapid desiccation of this region which compelled its inhabitants . . . pushing before them the former inhabitants of the lowlands to produce those great migrations and invasions of Europe which took place during the first centuries of our era.'

He went on to write that 'It is a geological epoch of desiccation that we are living in . . . nor is the phenomenon of desiccation limited to a small portion of the continent.'

The evidence for such progressive desiccation and the effects that it was held to have were not universally accepted. Sven Hedin (1940), the Swedish explorer, thought that much of the evidence for desiccation resulted from the shifting of river courses whilst Mackinder (1904) proposed that great leaders like Ghengiz Khan and Attila the Hun might be the cause of great invasions that created economic decline. Mill (1904) doubted that desiccation would be worldwide, and thought that desiccation in one place would be compensated for by widespread wetness elsewhere. Other workers, notably Sir Aurel Stein (1938), believed that the melting of relict Pleistocene glaciers and ice caps in the Himalayas would provide progressively less discharge to rivers during the course of the Holocene and so would promote the drying up of rivers and lakes without the direct reduction of precipitation envisaged by Kropotkin. Equally, in some areas groundwater reserves may have been filled during the pluvial phases, only to become gradually depleted during the post-glacial millennia. Lamb (1968) has recently proposed that such groundwater was 'a slowly wasting inheritance of the ice age . . . which may well have affected the habitability and cultivatability of the desert fringes until the time of Christ'.

Although these various points of view have weakened the simple Kropotkin hypothesis, the concept of post-glacial desiccation into historical times (even to the present) has been much followed in Indian literature of the 1950s and 1960s in spite of evidence to the contrary produced decades earlier by Blanford (1877) and Heron (1917). In particular, fears have been expressed that the Great Indian Sand Desert is on the move eastwards into eastern Rajasthan, Haryana, the Punjab and Uttar Pradesh, and plans have been formulated for the amelioration of the Rajasthan Desert (National Institute of Sciences, 1952). In 1960, D. N. Wadia, one of the most distinguished of Indian geologists, wrote (p. 1) that in many parts of Asia:

'. . . the same sequence of events has happened, increasing dryness, migration of the indigenous fauna and flora, erosion of the soil-cover by wind and undisciplined rush of water across the fields during the few occasional rain storms and the loss of vegetation cover. These ravages of nature have been supplemented by the acts of man. . . .'

He suggested (p. 6) that the Thar, before the Christian era, 'was a well watered and cultivated country' which supported flourishing Harappan (Indus civilization) cities and towns, but that 'during the last 20 centuries, the aridity curve on the whole has been a downward curve, the cause of which is to be sought in major atmospheric changes on a planetary scale' (p. 14). 'For long,' he said (p. 18), 'the deserts have carried on a winning fight against man and ousted large settled populations. . . .' Thus progressive desiccation was still in the 1960s a real issue in the study of both human and physical conditions in northern India.

The role of progressive desiccation was also explored with great vigour in West Africa, especially from 1920 onwards (see Monod, 1950; Worthington, 1958; and Prothero, 1962). In particular there was a fear expressed that from the Red Sea to the Atlantic coast the Sahara was encroaching upon the Sudan zone, that well levels were falling, that lakes were drying up, that rainfall was diminishing and that a general southward drift of people was taking place as a result of deteriorating conditions on the desert margins. Renner (1926, p. 587) remarked:

'There is a general belief that aridity is increasing in the Sudan. It is asserted that the water supply is diminishing: as a result of the gradually increasing aridity, wells are shrinking, lakes drying up, rivers ceasing to flow, and water holes filled up with sand. Certain crops are no longer grown, and pastures are being depleted. There seems to be a gradual southward movement of peoples throughout the region.'

This point of view had also been expressed by Bovill (1929) and by Hubert

(1920). It was championed in the 1930s by a forester with, significantly, Indian experience, called Stebbing (1935). He believed initially that the deterioration was essentially natural and remarked that 'the people are living on the edge, not of a volcano, but of a desert whose power is incalculable and whose silent and almost invisible approach must be difficult to estimate.' Hubert (1920), on the basis of an investigation of suspect long-term rainfall records for St Louis in Senegal, stated that the instrumental record suggested the current progress of desiccation.

The concern established by such opinions was such that an Anglo-French colonial forestry commission investigated the problem in the field in the late 1930s, but (see Jones, 1938) it failed to substantiate the gloomy Bovill–Hubert–Stebbing hypothesis. The Commission reported that whilst there was little evidence for climatic retrogression there was evidence for adverse human practices which produced dune sand reactivation and water-table lowering. Population movements towards the coast were regarded as being a response to economic forces. Moreover, workers such as Chudeau (1921) and Aubréville (1938), who reexamined the St Louis and Dakar rainfall records, did not find the steady and marked downward trend that Hubert supposed to exist.

In 1938 Jones pulled together a large quantity of material, especially geological, which effectively demolished many of the tenets of the progressive desiccation hypothesis as applied in West Africa. In particular he pointed out (as had Urvoy, 1935, and Falconer, 1911) that sand movement was less prevalent than at some phase in the past:

'Since anchored sand dunes occur in many of the old Quaternary river valleys it follows that the period of aridity intervened between the early Quaternary and the present day. Most of the evidence which has been advanced to support the theory of progressive desiccation is due to the regional desiccation during this arid period.'

He also recorded that Lake Chad was at a relatively high level in the 1930s whereas in the 1920s fears were expressed that it might dry up altogether (Tilho, 1928); Jones concluded that:

'West Africa has experienced several climatic changes since the beginning of the Quaternary period and the last major change appears to have been one towards more humid conditions . . . the climate has shown no great alteration during historic times . . . there is no need therefore to fear that desiccation *through climatic causes* will impair the habitability of the West African colonies for many generations to come.'

Stamp (1940) supported Jones and remarked that 'there now seems little doubt that the problem before West Africa is not the special one of Saharan encroachment, but the widespread one of man-induced soil-erosion . . .'

This belief in the role of soil erosion and impoverishment in causing some of the observed features of 'desertification' was substantiated by many French workers concerned with the spread of laterization and 'bovalisation' (Aubréville, 1947; Chevalier, 1950; Pelissier, 1951).

Probably the part of the world where the progressive desiccation hypothesis was held earliest was Southern Africa. As the agronomist Thompson (1936) has remarked:

> 'The opinion that South Africa is becoming more arid and that the desert is gradually encroaching is not new. Schwarz and others of the present generation who subscribe to this view are merely repeating the belief of generations ago. This popular notion has been handed down from one generation to another without any real proof.'

The explorers Lichtenstein (1803–6) and David Livingstone (1857) both supported the drying-up idea, though Livingstone, deeply influenced in his geological ideas by the catastrophist/structuralist notions of Sir Roderick Impey Murchison, thought that the desiccation was caused by the sudden draining of the Kalahari Lakes by the Zambesi. In 1865, Wilson, in turn influenced by Livingstone, drew a map of the areas being subjected, in his view, to increasing drought and he remarked:

> 'A very noticeable fact, which has of late years attracted considerable attention from residents in South Africa, is the gradual drying up of large tracts of country in the Trans 'Gariep. That great expanse of wilderness, called the Kalahari, remarkable for few inhabitants, little water and considerable vegetation seems to be gaining in extent, gradually swallowing up large portions of the habitable country on its confines, and slowly, but surely, assimilating their fertile character to its own sterile one.'

In 1875, Brown, in a study of South African hydrology, also suggested that desiccation was 'extreme' and 'still going on', and popular books referred to the 'Demon of the Desert' (Macdonald, 1914), but the most famous statement of the desiccation hypothesis, and one which had a great impact on both the public and the South African water authorities, came from Schwarz. He proposed (Schwarz, 1923) that the process could conceivably be reversed by deliberate flooding of the Kalahari depressions (such as Ngami, Makarikari and Mababe) by diversion of Cunene, Okavango and Zambesi waters, but warned:

> 'The droughts are becoming worse and worse every year, that people are even now becoming squeezed out of the central districts and are taking refuge on the coast, where conditions of life are still normal . . . the time is not far distant when the Karroo will become as desert as the Sahara.'

Soon substantiation for this pessimistic opinion came from Thompson

(1936) who analysed the available rainfall figures for South Africa and found that the pre-Boer War period was one of considerable wetness over a large part of that country. He then stated that:

'The fact cannot be denied that a diminution in rainfall has been taking place over a large part of South Africa during the last 40 to 50 years . . . it is in no way intended to imply that the country is actually drying up . . . and there is ample reason to believe that during some future epoch the present downward trend may be followed by an upward one.'

However, after a review of the evidence Kanthack (1930) rejected the idea that adverse climatic changes were taking place and suggested that deterioration of veld and water resources was the result of man-induced entrenchment of 'vlei' (valley bottom) alluvium. Moreover, a Kalahari Reconnaissance Commission (Department of Irrigation, Union of South Africa, 1926), sometimes called the Du Toit Commission, investigated the theories of Schwarz on an expedition around the Kalahari Sandveld in 1925, and reported unfavourably. The report of the Commission stated that (p. 8):

'the Professor's writings are characterised by an unfortunate tendency to expect his assertions to be accepted in the place of reasons, by amazing errors of idea, and by not a few inaccuracies of statement . . .'

and with regard to the question of desiccation in general stated:

'The idea of a former greater rainfall in the Ngami–Makarikari region in the historic period is based upon a misinterpretation of the facts and of the historic records, . . .'

and

'The climatic conditions do not appear to have altered appreciably within the limited period with which we are particularly concerned.'

However, Schwarz produced a large number of popular papers on the desiccation theme, with such titles as 'The lost lakes of South Africa' (1918) and 'The menace to Ovamboland' (1919), and promoted much discussion to his ideas (Schonken, 1924): 'I may say that there can be no travelled person present, who has not at some time or other personally encountered symptoms of an increase in aridity.'

THE HUMAN ROLE

As is evident from the above discussion of the concept of post-glacial progressive desiccation, there has also been a long-standing interest in the role that humans have played in degrading arid lands and their margins. For example, as a result of his consular experiences in the Middle East, the father

of conservation, George Perkins Marsh, was prompted to write his classic
Man and Nature (1864, pp. 9–10), in which he stressed the adverse environmental consequences of deforestation and other human actions:

> 'Besides the direct testimony of history to the ancient fertility of the
> regions to which I refer—Northern Africa, the Greater Arabian penin-
> sula, Syria, Mesopotamia, Armenia and many other provinces of Asia
> Minor, Greece, Sicily and parts of even Italy and Spain—the multitude
> and extent of yet remaining architectural ruins, and of decayed works of
> internal improvement, show that at former epochs a dense population
> inhabited those now lonely districts. Such a population could have been
> sustained only by a productiveness of soil of which we at present discover
> but slender traces . . . .
>
> The decay of these once flourishing countries is partly due no doubt to
> that class of geological causes, whose action we neither resist nor
> guide . . . , but it is in far greater proportion, either the result of man's
> ignorant disregard of the laws of nature, or an incidental consequence of
> war, and of civil and ecclesiastical tyranny and misuse.'

In the 1930s, especially as a result of the horrors of the Dust Bowl Years in
the American High Plains (see Worster, 1979, for a discussion of this
trauma), there was a rebirth of interest in land degradation as a problem
(Sauer, 1938), and Bennett (1938) produced a magisterial monograph on soil
erosion and conservation. A popular exemplification of the mood of this
period is *The Rape of the Earth* by Jacks and Whyte (1939, p. 63):

> 'Throughout Africa, and more so in the countries bordering on the
> Sahara, the problem of increasing desiccation (the so called "advance of
> the desert") and shortage of water supplies is becoming ever more acute.
> The deterioration of the forest through savanna to poor grassland and
> ultimately to bare ground has a disastrous effect upon the water relations
> of the many areas where it is occurring. The loss of equilibrium in the
> ecological balance means a disturbance in the hydrologic cycle particu-
> larly in the direction of a great increase in evaporation from the soil
> surface and a great decrease in the amounts added to the valuable
> underground reserves in the ground-water table. Thus, as in many similar
> parts of the world, the rainfall is not so much decreasing as becoming less
> effective, due to increased run-off and evaporation. The urgency of
> stopping the sinking of the ground-water level cannot be over-
> emphasized; all schemes of rehabilitation and revegetation will be of
> little avail if these resources are depleted.'

They doubt the importance of climatic changes but do not doubt that desert
expansion is occurring in an insidious and widespread fashion (pp. 170–1):

'The history of civilizations is a record of struggles against the progressive desiccation of agricultural land, but this land was reduced to a state of desolation and poverty by the hand of man more than by climatic change. This man-induced desiccation is still proceeding at the present day in that zone of delicate ecological balance between the humid and the true desert climates which occurs in many parts of the world. The deserts of Soviet Central Asia are threatening the irrigation systems of the fields and gardens which lie along the river valleys. The Sahara is being augmented by large areas of recently denuded and desiccated land on the south and south-west towards Nigeria, the Gold Coast and Sierra Leone, on the north-west to the foot of the High Atlas, on the north to Libya, and on the south-east towards Kenya and Uganda. The great Australian desert is being extended in the zones where it meets South Australia, New South Wales and south-western Queensland.'

Colonial governments took the spectre of a Dust Bowl in Africa very seriously and some conservation schemes were implemented (Anderson, 1984).

DESERTIFICATION

A new spasm of interest in the theme of desert expansion emerged in the late 1960s and 1970s as a result of two prime stimuli: the environmental revolution, and the Sahel drought and the Crisis in Africa. It was at this time that the terms 'desertization' and 'desertification' came into common parlance (Table 1.1), though there was a great deal of sterile terminological debate (see Verstraete, 1986, for a discussion). In some publications the very term was appropriated as the selling point of a book even though the contents were sometimes either more appropriate for a pure text in aeolian geomorphology (e.g. El-Baz and Hassan, 1986) or included material that had little reference to deserts (e.g. Fantechi and Margaris, 1986). In 1977, in Nairobi, the United Nations convened a Conference on Desertification (UNCOD), the purpose of which was to agree on a Plan of Action to combat desertification and bring it under control by the year 2000 AD.

The term 'desertification' was first used but not formally defined by Aubréville (1949), and for some years the term 'desertization' was also employed, as, for example, by Rapp (1974, p. 3), who defined it as: 'The spread of desert-like conditions in arid or semi-arid areas, due to man's influence or to climatic change.'

An alternative expression 'land aridization' has also been used by the Soviet pedologist, Kovda (1980, p. 15): 'The phrase "land aridization" means a complex of diverse processes and trends that reduce the effective moisture

Table 1.1   Selected books on desertification and related issues since 1970

Author/Editor	Date	Area
Glantz	1986	Worldwide
El-Baz and Hassan	1986	Worldwide
Rapp *et al.*	1976	Worldwide
Walls	1980	Worldwide
Kovda	1980	Worldwide
Grainger	1983	Worldwide
Dregne	1983	Worldwide
UNESCO/FAO	1977	Worldwide
Spooner and Mann	1982	Worldwide
Van Ypersele and Verstraete	1986	Worldwide
Chisholm and Dumsday	1987	Australia
Sabadell *et al.*	1982	USA
Sheridan	1981	USA
Fantechi and Margaris	1986	Europe
Glantz	1977	Sahel
Glantz	1987	Africa
Christiansson	1981	Tanzania
Biswas and Biswas	1980	Australia, China, Iran, Israel, USA, USSR

content over large areas and decrease the biological productivity of the soils and plants of an ecosystem.'

There has been some variability in how 'desertification' itself is defined. Some definitions stress the importance of human causes (e.g. Dregne, 1986, pp. 6–7):

'Desertification is the impoverishment of terrestrial ecosystems under the impact of man. It is the process of deterioration in these ecosystems that can be measured by reduced productivity of desirable plants, undesirable alterations in the biomass and the diversity of the micro and macro fauna and flora, accelerated soil deterioration, and increased hazards for human occupancy.'

Others admit the possible importance of climatic controls but give them a relatively inferior role (e.g. Sabadell *et al.*, 1982, p. 7):

'The sustained decline and/or destruction of the biological productivity of arid and semi arid lands caused by man made stresses, sometimes in conjunction with natural extreme events. Such stresses, if continued or unchecked, over the long term may lead to ecological degradation and ultimately to desert-like conditions.'

Yet others, more sensibly, are more even-handed or open-minded with respect to natural causes (e.g. Warren and Maizels, 1976, p. 1):

'A simple and graphic meaning of the word "desertification" is the development of desert like landscapes in areas which were once green. Its practical meaning . . . is a sustained decline in the yield of useful crops from a dry area accompanying certain kinds of environmental change, both natural and induced.'

Another controversy surrounds what should and what should not be incorporated within the term. This is brought out starkly when one compares the definitions of, first, Rapp (1987, p. 27) and then Mortimore (1987, pp. 2–3). Rapp defined desertification as:

'The spread of desert-like-conditions of low biological productivity to drylands outside the previous desert boundaries. Desertification is severe degradation of drylands, lasting more than one year, and is manifested by the loss of vegetation cover, loss of topsoil by wind or water erosion, reduction in primary productivity through soil exhaustion, salinization, or excessive deposition of sand dunes, sheets or coarse flood sediments.'

Mortimore, on the other hand, was more restrictive:

'The essence of desertification is "the diminution or destruction of the biological potential of the land" leading to the extension of desert-like conditions of soil and vegetation into areas outside the climatic desert, and the intensification of such conditions, over a period of time. I wish to exclude certain processes of ecological degradation which are commonly, but to my mind confusingly, included within the purview of the term. These are deforestation (which is the normal prelude to agricultural land use and is reversible), salinization of irrigated soils (which is caused by inadequate drainage), and soil erosion by water.'

Another restriction that is often placed on the use of the term is that it should comprise some notion of long-term and possibly irreversible or irreparable change (e.g. Wehmeier, 1980, p. 126):

'Desertification here will be understood as disadvantageous alterations, not oscillations, of and within ecosystems in arid and semi-arid areas. These alterations are usually triggered by man—involuntarily—, take place within a relatively short period of time—several years to several decades—, and very often cause irreparable or but partly repairable damage.'

The term 'desertification' is often confused with 'drought'. However, there is a difference. Drought is a relatively short-term problem which has acute phases lasting for a few years at a time. By contrast, desertification is a more chronic long-term problem. Drought does not directly result in desertification unless it is so long extended that an area is more or less permanently deprived of its precipitation. Normally, when the return of rain heralds the end of drought, vegetation returns. Desertification is also often confused with 'famine', but famine does not directly relate to the state of dryland environments but rather to food shortages, some of which may be caused by droughts.

The spatial character of desertification is also the subject of some controversy (Helldén, 1985). Contrary to popular rumour, the spread of desert-like conditions is not an advance over a broad front in the way that a wave overwhelms a beach. Rather it is like a 'rash' which tends to be localized around settlements. It has been likened to Dhobi's itch—a ticklish problem in difficult places (Goudie, 1981). Fundamentally, as Mabbutt (1985, p. 2) has explained, 'the extension of desert-like conditions tends to be achieved through a process of accretion from without, rather than through expansionary forces acting from within the deserts'. This distinction is important in that it influences perceptions of appropriate remedial or combative strategies.

There are relatively few reliable studies of the rate of supposed desert advance. Lamprey (1975) (see Figure 1.2) attempted to measure the shift of vegetation zones in the Sudan and concluded that the Sahara had advanced 90 to 100 km between 1958 and 1975, an average rate of about 5.5 km per year. However, on the basis of analysis of remotely sensed data and ground observation, Helldén (1984) found sparse evidence that this had in fact happened. One problem is that there may be very substantial fluctuations in biomass production from year to year. This has been revealed by meteorological satellite observations of green biomass production levels on the south side of the Sahara (Dregne and Tucker, 1988).

The 1977 Nairobi UNCOD meeting depicted on its maps four classes of desertification:

(a) Slight
  —little or no deterioration of plant cover or soil.
(b) Moderate
  —significant increase in undesirable forbs and shrubs or
  —hummocks, small dunes or small gullies formed by accelerated wind or water erosion or
  —soil salinity causing reduction in irrigated crop yields of 10–50%.
(c) Severe
  —undesirable forbs and shrubs dominate the flora,
  —sheet erosion by wind and water have largely denuded the land of vegetation, or large gullies are present or
  —salinity has reduced irrigated crop yields of more than 50%.
(d) Very severe
  —large shifting barren sand dunes have formed or
  —large, deep and numerous gullies are present or
  —salt crusts have developed on almost impermeable irrigated soils.

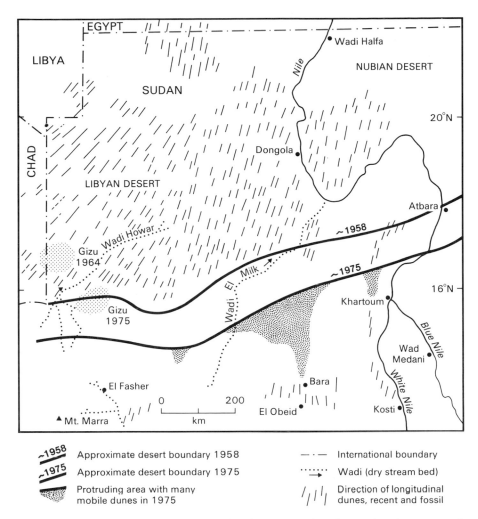

Figure 1.2 Map, derived from the work of Lamprey (1975) showing the supposed encroachment of desert in the northern Sudan between 1958 and 1975, as represented by the position of the boundary between sub-desert-scrub and grassland and the desert (after Rapp *et al.*, 1976, Fig. 8.5.3)

Building upon this, Mabbutt (1985) developed a threefold categorization of desertification in rangelands:

(a) Moderate
   —significant reduction in cover and deterioration in composition of pastures,
   —locally severely eroded,

Table 1.2 Desertification, early 1980s

	Total productive drylands		Productive dryland types					
			Rangelands		Rainfed croplands		Irrigated lands	
	Area (million hectares)	Percent desertified	Area (million hectares)	Percent desertified	Area (million hectares)	Percent desertified	Area (million hectares)	Percent desertified
Total	3257	61	2556	62	570	60	131	30
Sudano–Sahelian Africa	473	88	380	90	90	80	3	30
Southern Africa	304	80	250	80	52	80	2	30
Mediterranean Africa	101	83	80	85	20	75	1	40
Western Asia	142	82	116	85	18	85	8	40
Southern Asia	359	70	150	85	150	70	59	35
USSR in Asia	298	55	250	60	40	30	8	25
China and Mongolia	315	69	300	70	5	60	10	30
Australia	491	23	450	22	39	30	2	19
Mediterranean Europe	76	39	30	30	40	32	6	25
South America and Mexico	293	71	250	72	31	77	12	33
North America	405	40	300	42	85	39	20	20

Source: UN Environment Program.

—would respond to management supported by improvements and conservation measures,
—loss of carrying capacity up to 25% of earlier carrying capacity.
(b) Severe
—very significant reduction in perennial vegetation cover and widespread deterioration in composition of pastures,
—widespread severe erosion,
—requiring major improvements,
—loss of carrying capacity 25–50% of earlier carrying capacity.
(c) Very severe
—extensively denuded of perennial shrubs and grasses and subject to
widespread very severe accelerated erosion,
—large areas irreclaimable economically,
—loss of carrying capacity over 50 per cent of earlier carrying capacity.

Using data based on these categories he estimated that on a global basis 4500
million hectares were at risk from desertification and that 945 million hectares
were severely or very severely desertified.

Table 1.2 presents UNEP data for different regions as to the amount of
desertification that has occurred in the 3275 million hectares of 'productive'
drylands. It is estimated that 61% showed evidence of desertification, with
the greatest percentage being in Sudano–Sahelian Africa (about 88%). In
all UNEP has estimated that desertification threatens 35% of the earth's land
surface (about 45 million km^2) and 19% of its population (some 850 million
people) (Stiles, 1984).

HUMAN CAUSES OF DESERT DEGRADATION

The fundamental cause of most problems of arid zone degradation is the
twentieth century explosion in human population levels. Table 1.3 presents
data for selected countries with substantial areas of arid zones in Africa and
Asia. The increase over the period 1900–80 ranges from just under ×3 to just
under ×7.

This has led to four main direct causes of desertification: overcultivation,
overgrazing, deforestation and salination of irrigation systems. We will now
consider these causes, and other subsidiary ones, in turn.

Overcultivation

A fundamental cause of arid zone degradation is overcultivation and the
creep of dryland agriculture into excessively dry areas. Cropping now takes
place in areas receiving as little as 150 mm of annual rainfall in North Africa
and the Near East and 250 mm in the Sahel (Le Houérou, 1977, p. 26). Soil

Table 1.3   Changing human populations (in millions) in the twentieth century

Country	Year					Percent increase 1900–80
	1900	1925	1950	1975	1980	
Algeria	5	6	9	16	19	380
Botswana	0.12	0.175	0.31	0.66	0.82	683
Egypt	10	14	20	37	42	420
India	237	260	356	600	664	280
Morocco	5	6	9.5	17.5	20.2	404
Pakistan	16	22	33	70	81.5	509
Sahel States (Mauritania, Chad, Mali, and Niger)	6	6.5	8.5	15	18.3	305
Sudan	6	7	9	13	18.7	317

Source: Data in McEvedy and Jones (1978) and UN Statistics.

left barren after cropping or after crop failure is prone to wind and water erosion. Especially prone to the former are areas of relict late Pleistocene dune fields (e.g. Rajasthan, Sahel, The Mallee of South Australia, etc.).

Excessive human population levels may build up in particular areas because of political direction, and this can lead to excessive pressures on land. One of the more extreme examples of this is provided by Namibia (South-West Africa), where influx controls mean that most of the black population lives in a narrow northern strip, with the remainder of the country being largely given over to extensive white ranches, game reserves, etc. (Wellington, 1967).

Overgrazing

Overgrazing is widely regarded as a prime cause of desertification. Globally the population of cattle rose by 38% between 1955 and 1976, and that of sheep and goats by 21%. In many areas the increases in free-ranging livestock populations have exceeded the carrying capacity of the land. Furthermore, in some regions increases in the area under cultivation have reduced available pastureland and intensified pressure on remaining pastures. Some animals may be more prone to create desertification than others. Sheep, for example, are both highly gregarious and are generally incapable of moving long distances from waterholes, so that pasture abuse may be severe. At the other end of the spectrum, camels have certain advantages in the fight against desertification in comparison with cattle. They produce more milk per lac-

Table 1.4 Comparison of values of camel and cattle pastoralism

	Camels	Cattle
Annual milk production for human use, one cow (litres)	1300–2500	112–420
Lactation period (weeks)	47–72	16–60
Herd of 100, annual production		
milk (kg)	24 820	6615
blood (kg)	356	480
meat (kg)	675+	960
total protein (kg)	1100	410
total energy (kcal)	18 730 000	7 882 500
Dry matter intake (kg/head)		
per day	10	7.5
per year	3650	2737.5
Typical herd annual growth rate (%)	1.5	3.4
Maximum herd annual growth rate (%)	7.5	15.0
Herd size necessary to sustain an average family (6 people)	28	64
Diet	Trees and shrubs (70%)	Grass (80%)
Mobility	High	Moderate
Trampling affects	Light	Heavy
Pastoralist degree of polygyny	Low	High
Human population growth	Low	High

From various sources in Stiles, 1983, Table 1.

tation, the milk-giving period is longer and the camel continues to produce adequately throughout the dry season. They have more varied diets than cows, can travel further in a day (causing a lower intensity of grazing and trampling round settlements), are less dependent on watering places (and so can exploit a much larger proportion of the available range) and are more efficient than the cow in terms of vegetation consumed for milk produced (Table 1.4).

The tendency for some introduced domestic species to become feral may have severe ecological implications. The classic case of this is in Australia where rabbits, feral cattle, donkeys, horses and goats are present in large numbers in certain areas. They contribute to the overgrazing problem (Noble and Tongway, 1986).

Although domesticated animals are the prime cause of vegetation degeneration, there are specific circumstances where wild animals, especially ungulates, can create desertification. In the semi-arid and sub-humid low veld of Swaziland, for example, fast-breeding impala have severely degraded the

Hlane and Mlaula game reserves, because they are not exposed to natural predation or adequate culling. Areas with nutritious grasses (e.g. on basalt soils and alluvium) suffer far more than areas on poorer substrata (e.g. on acid volcanics and sandstones).

Pastoral nomads, though sometimes labelled as 'backward' and 'conservative', had in many cases developed various types of adaptation which enabled them to survive in, and in many cases maintain the utility of, marginal areas. However, in many areas the traditional systems have broken down, so that the equilibrium between people and land has become disrupted. Migrations, for example, have been curbed by the establishment of national boundaries or by deliberate government policies of sedentarization, while the expansion of cash crop cultivation may force nomads to use smaller and smaller areas of grazing, with resulting overgrazing effects.

The installation of modern boreholes and various types of excavated waterhole has enabled rapid multiplication of livestock numbers, so that excessive overgrazing and land degradation occurs in the proximity of the new water sources. Without the rest period that was previously assured by intermittent water supplies, vegetation replaced water as the limiting factor in livestock survival.

Land use changes (e.g. changes in grazing pressure, burning, etc.) can lead to bush encroachment—the invasion of productive grassland by tree and shrub species which are frequently unpalatable and of little economic use.

Deforestation

The uprooting of woody species is a further fundamental cause of desert degradation. As Le Houérou (1977, p. 27) put it:

'Minimal wood consumption for domestic uses is about one kg per person per day, and often it is more than three times that figure. Given that the average woody biomass of a steppe in good condition, either in the Mediterranean or in the tropics, is between 500 and 1000 kg per hectare, including the main roots, each person dependent on this type of fuel would destroy at least half a hectare every year if there were no regeneration. If we assume that there is some regeneration on the average of every second year each family of five persons would still destroy more than one hectare of woody steppe per year. As over 100 million people depend on this type of fuel for their daily needs in the arid zones of Africa and the Middle East, we can estimate a theoretical destruction of about 25 million hectares per year. This is, of course, mere speculation but it shows the magnitude of a very serious cause of desertisation which is often overlooked.'

The collection of wood for charcoal and firewood is an especially serious problem in the vicinity of large urban areas where electricity and electrical devices (e.g. stoves) are too expensive for most of the urban poor. Chidumayo (1983), for example, has catalogued the consequences of the elevenfold increase in the population of Lusaka, Zambia, between 1945 and 1980. Indeed it can be argued that it is the rising demand for firewood by urban dwellers, rather than the needs of rural people, that is frequently the cause of the problem. Acacia woodlands, formerly extensive, have disappeared within a 100 km radius of Khartoum, and land within 40 km of Kano (Nigeria) and Ougadougou (Burkina Faso) has been stripped of trees. Fuelwood and charcoal have to be transported to cities over ever-increasing distances—up to 500 km for charcoal to supply Dakar and Khartoum. The importance of tree conservation is discussed in Chapter 7.

Salinization

Humanly induced salinization resulting from the spread of irrigation is a sinister and widespread form of desert degradation (Table 1.5). As Jacobsen and Adams (1958) demonstrated for Mesopotamia, it is also a long-established problem, having caused serious problems by 4000 years ago. Salinization destroys soil structure, kills plants and reduces plant growth. It is also frequently accompanied by waterlogging. In Pakistan no less than 9.5 million hectares are adversely affected by salinity and waterlogging (Gazdar, 1987). This is a theme discussed in Chapter 4.

Salinization can result from vegetation clearance, for removal of native forest of bush vegetation allows a greater penetration of rainfall into deeper soil layers which causes groundwater levels to rise, creating seepage of

Table 1.5 Estimates of percentage of irrigated land affected by salinization for selected countries

Algeria	10–15	India	27
Egypt	30–40	Iran	<30
Senegal	10–15	Iraq	50
Sudan	<20	Israel	13
United States	20–5	Jordan	16
Colombia	20	Pakistan	<40
Peru	12	Sri Lanka	13
China	15	Syrian Arab Republic	30–5
Cyprus	25	Australia	15–20

Source: Data in Table 19.3 of *World Resources 1987*, a report by the International Institute of Environment and Development and the World Resources Institute, published by Basic Books, Inc., New York.

sometimes saline water in low-lying areas. This is a serious problem in the wheat belt of Western Australia and in the prairies of Canada and North Dakota (Peck, 1978).

In coastal areas overpumping of aquifers causes sea-water incursion. Fresh water is rapidly replaced by salt. A comparable situation can arise in coastal deltas (e.g. the Nile) if upstream damming of a river reduces the flow of fresh water to the delta.

Miscellaneous Causes and Symptoms

Soil compaction and crusting   Heavy agricultural machinery and excessive trampling by livestock may cause soil compaction which in turn lowers crop yields, decreases the rate of waste infiltration and increases runoff and erosion. Surface crusting, promoted by exposing the soil to the impact of raindrop impact through inappropriate cultivation practices and overgrazing, also increases runoff and erosion and interferes with seedling emergence.

Dung use for fuel   Firewood scarcity leads to the use of cattle dung for fuel, especially in the Indian sub-continent. In India around 300–400 million tons of wet dung (60–80 million tons dry weight) is burned for fuel each year. This represents a severe loss of potential soil nutrients, but also leads to a reduction in the quality of soil structures (Eckholm, 1977).

Eucalyptus and decline of water resources   From time to time the planting of species of *Eucalyptus* has been blamed for declines in water resources, both in terms of moisture and groundwater. Particular fears have been expressed in India (Shiva and Bandyopadhyay, 1986).

Depletion of groundwater resources   The excessive exploitation or 'mining' of groundwater reserves can cause severe decline in the levels of freshwater aquifers. This can eventually make the cost of pumping prohibitive. It can also cause loss of habitats, desiccation of oases and ground subsidence. One long-term consequence of groundwater overdraft and subsidence is that as an aquifer system compresses with the mining of its water, the amount of pore space within it shrinks, thereby greatly diminishing its storage capacity (Sheridan, 1981, p. 51).

Lake desiccation as a result of interbasin water transfers and water obstruction
The diversion of river water by interbasin water transfers or the diminution of discharge as a consequence of water use for irrigation and other purposes can lead to catastrophic declines in the levels of inland seas and lakes in arid areas. This can have a widespread series of adverse environmental impacts, including elevated salinity levels, decline of riparian vegetation and dispersal of saline dust by dust-storms. The most severe example of this type is the change in the status of the Aral Sea (Micklin, 1988) in the Soviet Union.

Damage resulting from vehicular access   Fragile desert ecosystems and susceptible soils may be prone to damage as a result of uncontrolled vehicular traffic (e.g. by recreational off-road vehicles, seismic parties, farmers).

Mining   In some arid areas mining may produce elements of desertification. In Australia, Mabbutt (1986, p. 107) mentions *inter alia*:

(a) destruction of vegetation and surface disturbance by tracks, pipelines, excavations, etc.,
(b) removal of trees and shrubs for timber and fuel,
(c) occurrence and sand drifting and dust from mine dumps,
(d) pollution of limited freshwater resources.

Loss of land because of labour shortage   In some areas (e.g. North Yemen) agriculture had traditionally depended on labour-intensive maintenance of soil conservation structures such as terracing, but emigration of a large section of the labour force to work in the lucrative economies of oil-rich neighbours has meant that the labour is no longer available to maintain the terraces (Alkämper *et al.*, 1979; Kassas, 1987).

Abandonment of rotation and gum arabic cultivation   In parts of the Sudan traditional shifting agriculture is divided into two phases making up a ten-year rotation which is ecologically sound. In the first stage land is cleared and cultivated for four or so years but is then abandoned as yields fall and fields become infested with weeds. Secondary tree growth occurs and is dominated by *Acacia senegal*, which is harvested for gum arabic. After 6–10 years of explloutation the trees die and fall down, providing protection against grazing so that dense grass takes over. Under population pressure the fallow periods are being reduced (Kassas, 1987).

Sedimentation   Accelerated erosion in one locality causes accelerated sedimentation at another, and sedimentation can have a whole suite of adverse environmental impacts including water pollution, infilling of reservoirs, smothering of plants and increasing flooding.

Loss of land to urbanization   Urbanization has taken place at a greater rate than the rate of world population growth as a whole. From 1965 to 1975, whereas world population grew at an annual rate of 1.9%, world urban population growth was 3.2%, and the annual growth of dryland cities was approximately 3.9% (Cooke *et al.*, 1982). Some dryland cities grew much faster than this, causing severe consumption of land.

Vegetation change because of wildlife decline   It is possible that in those areas where human predation has led to a great diminution in wildlife there may have been consequential changes in vegetation as the roles of wildlife in plant stimulation through browsing and in seed dispersal are reduced. This may have been especially important in relation to the regeneration of large-seed legumes such as *Acacia, Albizia, Bauhinia, Cassia, Entada, Parkia, Prosopis tetrapleura* and related genera (IUCN, 1986, p. 25).

THE QUESTION OF HUMAN-INDUCED CLIMATIC DETERIORATION

It has often been proposed that human activities are capable of creating adverse climatic changes in arid areas, and various mechanisms have been postulated.

The Charney Model of Albedo Change

Charney (1975) and Charney *et al.* (1977) have used a general circulation model (GCM) to show that increasing the albedo north of the Intertropical Convergence Zone (ITCZ) from 14 to 35% would have the effect of shifting the ITCZ southwards by several degrees of latitude, so decreasing the rainfall in the Sahel summer by about 40%. The albedo change could be produced by deforestation and overgrazing. The mechanism of this effect is that a higher albedo leads to less absorption of solar radiation by the ground, which causes less sensible and latent heat transfer to the atmosphere, which in turn produces less rain-giving convective clouds.

The Walker and Rowntree Soil Moisture Model

Walker and Rowntree (1977) demonstrated that ground dryness can cause deserts to persist, because rain-giving depressions cannot be nourished by the evaporation of soil moisture. Thus changes in surface albedo could have a positive feedback effect which is enhanced by this process.

The Mitchell Carbon Dioxide Model

The 'greenhouse effect' caused by increasing levels of anthropogenically released carbon dioxide (CO_2) may be the cause of increasing global temperatures for much of the last century, and such changes in temperature could modify atmospheric circulation patterns and the location and magnitude of precipitation. Modelling by Mitchell (1983) suggests that higher CO_2 levels and global temperatures would in general tend to even lower precipitation levels in areas that are already arid.

The Bryson and Barreis Dust Model

An increase in dust levels of the atmosphere, caused by wind deflation of susceptible surface materials following on from vegetation removal or surface disturbance, would serve to modify the scattering and absorption of solar radiation in the atmosphere. Convective activity might become reduced (Bryson and Barreis, 1967).

These four models are as yet largely speculative, and one test of their applicability is provided by an analysis of whether or not clear trends of climatic change are actually occurring in arid zones today.

CLIMATIC DETERIORATION: FACT OR FALLACY?

Implicit in many of the definitions of desertification and in the early arguments about desiccation is the belief that climatic deterioration may contribute to desert degradation.

When one examines recent rainfall data for the arid areas of the Sudan–Sahel in Africa, central Australia, north-west India and Arizona, USA (Figure 1.3), it is apparent that some areas show relatively little evidence of a downward trend in the last three or four decades, whereas others do.

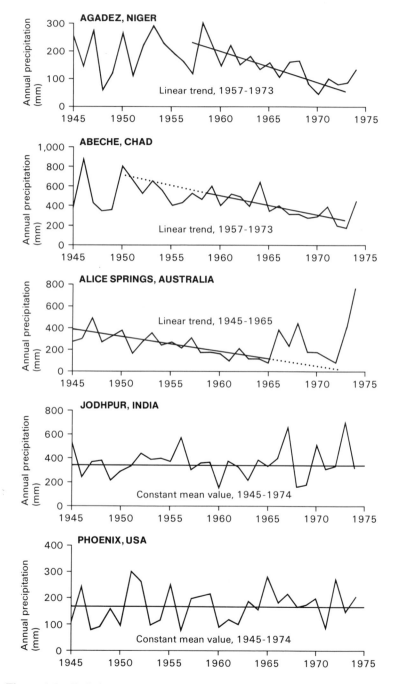

Figure 1.3   Rainfall variations at selected arid zone stations since 1945

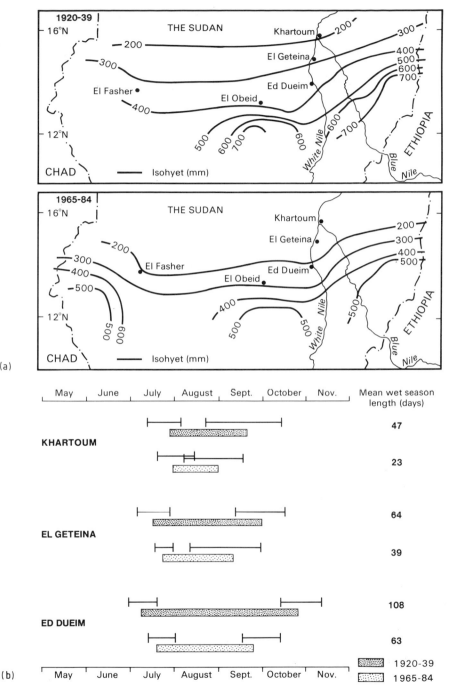

Figure 1.4 (a) The shifting position of the mean annual rainfall in the Central Sudan between 1920 and 1939 and 1965 and 1984 (from Walsh *et al.*, 1988, Fig. 4). (b) Changes in the duration of the wet season at three locations in Central Sudan over the same period. The bars indicate the interquartile ranges of wet season onset and termination dates (from Walsh *et al.*, 1988, Fig. 5)

Recent climatic deterioration has certainly been severe in the Sudan (Figure 1.4) where, in White Nile Province, annual rainfall in 1965–84 was 40% below 1920–39 levels, and wet season length contracted by 39–51% (Walsh *et al.*, 1988). The dry epoch which started in the mid 1960s has continued and intensified in the 1980s. Further west, the dry epoch has had dramatic effects on Lake Chad. Its area declined from 23 500 km^2 in 1963 to about 2000 km^2 in 1985, which is probably the lowest level of the century (Rasmusson, 1987, p. 156). A recent appraisal of rainfall trends in the Sahel is provided by Dennett *et al.* (1985) and for the western Sudan by Eldredge *et al.* (1988). Both studies agree on the existence of a clear downward tendency since the mid 1960s.

A major consequence of this rainfall decline on the south side of the Sahara was a greatly increased incidence of dust-storm activity (Middleton, 1985).

In southern Africa, by contrast, analyses of long-term meteorological records fail to show any progressive downward trend (Tyson, 1986). More evident is an 18 year cyclic pattern in areas of summer rainfall.

In the Rajasthan Desert of north-west India the trend of rainfall in the twentieth century appears to be very different from that in the Sahel. The latest analyses of monsoonal summer rainfall for the Rajasthan desert (Pant and Hingane, 1988) indicate that there has been a modest upward trend in precipitation levels between 1901 and 1982 (Figure 1.5).

In the drought-prone region of north-east Brazil, Hastenrath *et al.* (1984) have undertaken an analysis of rainfall records since 1912 (Figure 1.6). The incidence of runs of dry years is thereby highlighted, but there is no very conspicuous evidence of any long-term trend, either upwards or downwards.

In Australia, Hobbs (1988) provides an up-to-date analysis of rainfall trends, and data for Western and South Australia are presented in Figure 1.7. There is no clear-cut trend comparable to that found in the Sudan and Sahel belts of Africa. However, some trends are evident, though they vary in direction across the continent. Hobbs concludes (p. 295):

'The picture of variability for Australia is complex in both time and space, but this is not unexpected in view of the size of the continent. The mediterranean climatic regions [South Australia and Western Australia] . . . both show considerable rainfall variability on apparently irregular time scales. The major variations in the two regions have been out of phase with each other . . . . The evidence for any sustained long-term climatic changes, at least as far as rainfall is concerned, is unclear.'

From the point of view of erosion and landscape change, the variability in mean annual rainfall totals of the type discussed this far may be less relevant than changes in the incidence of such parameters as rainfall intensity. This was a theme which Cooke and Reeves (1976) investigated with respect to the development of erosional phases in the bottomlands of California and Ari-

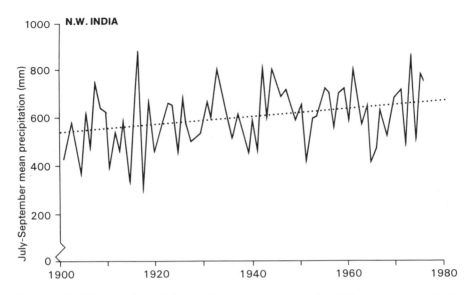

Figure 1.5 The trend of July to September mean rainfall for north-west India (modified after Pant and Hingane, 1988)

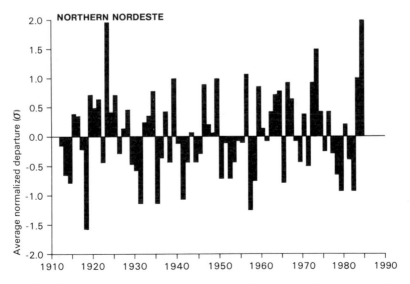

Figure 1.6 The standardized March–April rainfall series for the northern Nordeste of Brazil (from the work of Hastenrath, in Ward *et al.*, 1988, Fig. 21.2)

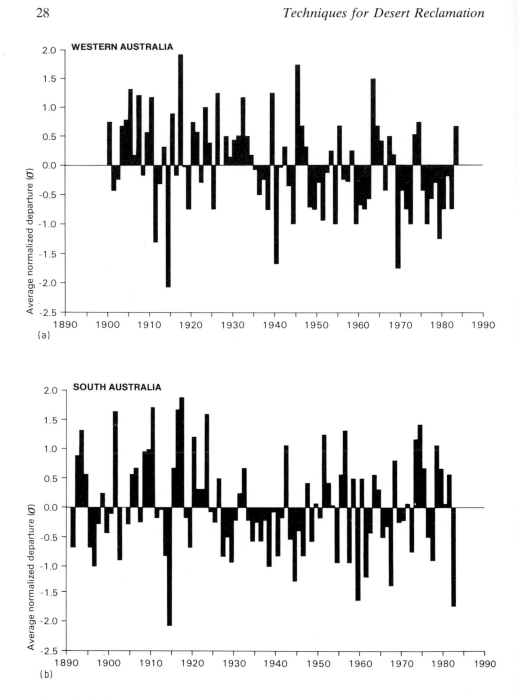

Figure 1.7   Time series of normalized winter rainfall departures for: (a) Western
Australia and (b) South Australia (from Hobbs, 1988, Fig. 25.2)

zona. In Arizona, the long-term rainfall record for Fort Lowell-Tucson, while showing no significant secular trends in annual precipitation totals from the late 1860s, did show an increasing incidence of low magnitude rain events and a decreasing incidence of high magnitude events. The lack of trend in mean annual rainfall in Arizona is demonstrated by the plot for Phoenix in Figure 1.3 and has also been demonstrated for California for the period 1850–1977 by Granger (1979).

In the rest of this volume we shall investigate some of the manifestations of desert degradation and consider some of the technological solutions to these various manifestations.

REFERENCES

Alkämper, J., Haffner, W., Mater, H. E. and Weise, O. R. (1979). *Erosion Control and Afforestation in Haraz, Yemen Arab Republic*, Giessen, Tropeninstitut.

Anderson, D. M. (1984). Depression, Dust Bowl, demography and drought: the colonial state and soil conservation in East Africa during the 1930s, *African Affairs*, **83**, 321–43.

Aubréville, A. (1938). La forêt coloniale: les forêts de l'Afrique occidentale française, *Annales, Academie des Sciences Coloniales*, **9**, 244.

Aubréville, A. (1947). 'Erosion et 'bovalisation' en Afrique noire française', *L'Agronomie Tropicale*, **2**, 339–57.

Aubréville, A. (1949). *Climats, Forêts et Desertification de L'Afrique Tropicale*, p. 351, Societé d'Edition Géographiques Maritimes et Coloniales, Paris.

Bennett, H. H. (1938). *Soil Conservation*, McGraw-Hill, New York.

Biswas, M. R. and Biswas, A. K. (1980). *Desertification*, p. 523, Pergamon Press, Oxford.

Blanford, W. T. (1877). Geological notes on the great Indian Desert between Sind and Rajputana, *Records, Geological Survey of India*, **10**, 10–21.

Bovill, E. W. (1929). The Sahara, *Antiquity*, **3**, 4–23.

Bowman, I. (1935). Our expanding and contracting 'deserts', *Geographical Review*, **25**, 43–61.

Brooks, C. E. P. (1922). *The Evolution of Climate*, London.

Brown, J. C. (1875). *Hydrology of South Africa or Details of the Former Hydrological Conditions of the Cape of Good Hope and of Causes of Its Present Aridity*, p. 260, Kirkcaldy.

Bryson, R. A. and Barreis, D. A. (1967). Possibility of major climatic modifications and their implications: northwest India, a case for study, *Bulletin of the American Meteorological Society*, **48**, 136–42.

Charney, J. G. (1975). Dynamics of deserts and drought in the Sahel, *Quarterly Journal of the Royal Meteorological Society*, **101**, 193–202.

Charney, J., Quirk, W. J. Chow, S. H. and Kornfield, J. (1977). A comparative study of the effects of albedo change on drought in the semi-arid regions, *Journal of Atmospheric Science*, **34**, 1366–85.

Chevalier, A. (1950). La progression de l'aridité, du dessechement et de l'ensablement et la décadence des sols en Afrique occidentale française, *Comptes rendus Academie des Sciences de Paris*, **230**, 1530–3.

Chidumayo, E. N. (1983). Urbanisation and deforestation in Zambia, *Desertification Control Bulletin*, **9**, 40–3.

Chisholm, A. and Dumsday, R. (Eds.) (1987). *Land Degradation Problems and Policies*, p. 404, Cambridge University Press, Cambridge.

Christiansson, C. (1981). *Soil Erosion and Sedimentation in Semi-arid Tanzania*, p. 208, Scandinavian Institute of African Studies, Uppsala.

Chudeau, R. (1921). La probleme du dessèchement en Afrique Occidentale, *Bulletin Comité d'Etudes Historiques et Scientifiques de l'A.O.F.*, pp. 353–69.

Coching, C. (1926). Climatic pulsations during historic times in China, *Geographical Review*, **16**, 274–82.

Cooke, R. U. and Reeves, R. W. (1976). *Arroyos and Environmental Change in the American south-west*, p. 213, Clarendon Press, Oxford.

Cooke, R. U., Brunsden, D., Doornkamp, J. C. and Jones, D. K. C. (1982). *Urban Geomorphology in Drylands*, p. 324, Oxford University Press, Oxford.

Curry, J. C. (1928). Climate and migrations, *Antiquity*, **2**, 292–307.

Dennett, M. D., Elston, J. and Rodgers, J. A. (1985). A reappraisal of rainfall trends in the Sahel, *Journal of Climatology*, **5**, 353–61.

Department of Irrigation, Union of South Africa (1926). *Report of the Kalahari Reconnaissance of 1925*, Government Printer, Pretoria.

Dregne, H. E. (1983). *Desertification of Arid Lands*, Harwood, Chur.

Dregne, H. E. (1986). Desertification of arid lands, in *Physics of Desertification* (Eds. F. El-Baz and M. H. A. Hassan), pp. 4–34, Nijhoff, Dordrecht.

Dregne, H. E. and Tucker, C. J. (1988). Desert encroachment, *Desertification Control Bulletin*, **16**, 16–19.

Eckholm, E. P. (1977). The other energy crisis, in *Desertification* (Ed. M. H. Glantz), pp. 39–56, Westview Press, Boulder.

El-Baz, F. and Hassan, M. H. A. (1986). *Physics of Desertification*, p. 473, Martinus Nijhoff, Dordrecht.

Eldredge, S., El Sayeed Khalil, S., Nicholds, N., Ali Abdaua, A. and Rydjeski, D. (1988). Changing rainfall patterns in western Sudan, *Journal of Climatology*, **8**, 45–53.

Falconer, J. D. (1911). *Geology and Geography of Northern Nigeria*, Macmillan, London.

Fantechi, R. and Margaris, N. S. (1986). *Desertification in Europe*, p. 231, Reidel, Dordrecht.

Gazdar, M. N. (1987). *Environmental Crisis in Pakistan*, p. 62, The Open Press, Kuala Lumpur.

Glantz, M. H. (Ed.) (1977). *Desertification*, Westview Press, Boulder.

Glantz, M. H. (Ed.) (1986). *Arid Land Development and the Combat against Desertification: An Integrated Approach*, p. 145, Centre of International Projects GKNT, Moscow.

Glantz, M. H. (Ed.) (1987). *Drought and Hunger in Africa*, p. 487, Cambridge University Press, Cambridge.

Goudie, A. S. (1972). The concept of post-glacial progressive desiccation, School of Geography, University of Oxford, Research Paper 4, 48 pp.

Goudie, A. S. (1981). Desertification, in *The Dictionary of Physical Geography* (Ed. R. J. Johnston), p. 77, Blackwell, Oxford.

Grainger, A. (1983). *Desertification*, p. 94, Earthscan, London.

Granger, O. (1979). Increased variability in California precipitation, *Annals Association of American Geographers*, **69**, 533–43.

Hastenrath, S., Ming-Chin, W. and Pao-Shin, C. (1984). Toward the monitoring and prediction of north-east Brazil droughts, *Quarterly Journal of the Royal Meteorological Society*, **118**, 411–25.

Hedin, S. (1940). *The Wandering Lake*, Routledge, New York.

Helldén, U. (1984). Drought impact monitoring, Lunds Universitets Naturgeografiska Institution, Rapporter och Notiser 61, Lund, Sweden, 61 pp.

Helldén, U. (1985). Land degradation and land productivity monitoring—needs for an integrated approach, in *Land Management and Survival* (Ed. A. Hjort), pp. 77–87, Scandinavian Institute of African Studies, Uppsala.

Heron, A. M. (1917). The geology of north-eastern Rajputana and adjacent districts, *Memoirs Geological Survey of India*,45 (1), 128.

Hobbs, J. E. (1988). Recent climatic change in Australasia, in *Recent Climatic Change* (Ed. S. Gregory), pp. 285–97, Belhaven Press, London.

Hubert, H. (1920). Le desséchement progressive en Afrique Occidentale francaise, *Bulletin Comité d'études Historiques et Scientifiques de l'Afrique Occidentale Française*, **1920**, 401–37.

IUCN (1986). *The IUCN Sahel Report*, p. 80, ICUN, Gland.

Jacks, G. V. and Whyte, R. O. (1939). *The Rape of the Earth: A World Survey of Soil Erosion*, p. 313, Faber and Faber, London.

Jacobsen, T. and Adams, R. M. (1958). Salt and silt in ancient Mesopotamian agriculture, *Science*, **128**, 1251–8.

Jones, B. (1938). Desiccation and the West African colonies, *Geographical Journal*, **91**, 401–23.

Kanthack, F. E. (1930). The alleged desiccation of South Africa, *Geographical Journal*, **76**, 516–21.

Kassas, M. (1987). Seven paths to desertification, *Desertification Control Bulletin*, **15**, 24–6.

Kovda, V. A. (1980). *Land Aridization and Drought Control*, p. 277, Westview Press, Boulder.

Kropotkin, P. (1904). The desiccation of Eur-Asia, *Geographical Journal*, **23**, 722–41.

Lamb, H. H. (1968). The climatic background to the birth of civilisation, *Advancement of Science*, **25**, 103–20.

Lamprey, H. F. (1975). *Report on the Desert Encroachment Reconnaissance in Northern Sudan*, p. 16, Ministry of Agriculture, Khartoum.

Le Houérou, H. N. (1977). The nature and causes of desertization, in *Desertification* (Ed. M. H. Glantz), pp. 17–38, Westview Press, Boulder.

Lichtenstein, H. (1803–6). *Travels in Southern Africa*, Van Riebeck Society, Cape Town.

Livingstone, D. (1857). *Missionary Travels and Researches in South Africa*, Murray, London.

Mabbutt, J. A. (1985). Desertification of the world's rangelands, *Desertification Control Bulletin*, **12**, 1–11.

Mabbutt, J. A. (1986). Desertification in Australia, in *Arid Land Development and the Combat Against Desertification: An Integrated Approach*, pp. 101–12, UNEP, Moscow.

Macdonald, W. (1914). *The Conquest of the Desert*, p. 197, Laurie, London.

McEvedy, C. and Jones, R. (1978). *Atlas of World Population History*, p. 368, Penguin, Harmondsworth.

Mackinder, H. J. (1904). Discussion, *Geographical Journal*, **23**, 734–6.

Marsh, G. P. (1864). *Man and Nature*, Scribner, New York.

Micklin, P. P. (1988). Desiccation of the Aral Sea: a water management disaster in the Soviet Union, *Science*, **241**, 1170–6.

Middleton, N. J. (1985). Effect of drought on dust production in the Sahel, *Nature*, **316**, 431–4.

Mill, H. R. (1904). Discussion, *Geographical Journal*, **23**, 739–40.
Mitchell, J. F. B. (1983). The seasonal response of a general circulation model to changes in CO_2 and sea temperatures, *Quarterly Journal of the Royal Meteorological Society*, **109**, 113–52.
Monod, T. (1950). Autour du problème du desséchement Africain, *Bulletin Institut Français Afrique Noire*, **12**, 514–23.
Mortimore, M. (1987). Shifting sands and human sorrow: social response to drought and desertification, *Desertification Control Bulletin*, **14**, 1–14.
National Institute of Sciences (1952). *Proceedings of the Symposium on the Rajputana Desert*, Bulletin No. 1.
Noble, J. C. and Tongway, D. J. (1987). Herbivores in arid and semi-arid rangelands, in *Australian Soils; The Human Impact* (Eds. J. S. Russell and R. F. Isbell), pp. 243–70, University of Queensland Press, St Lucia.
Pant, G. B. and Hingane, L. S. (1988). Climatic change in and around the Rajasthan Desert during the twentieth century, *Journal of Climatology*, **8**, 391–401.
Peck, A. J. (1978). Salinization of non-irrigated soils and associated streams: a review, *Australian Journal of Soil Research*, **16**, 157–68.
Pelissier, P. (1951). Sur la desertification des territoires septentrionaux L'A.O.F., *Cahiers d'outre Mer*, **4**, 80–5.
Prothero, J. M. (1962). Some observations on desiccation in north-western Nigeria, *Erdkunde*, **16**, 111–19.
Rapp, A. (1974). A review of desertization in Africa—water, vegetation and man, Secretariat for International Ecology, Sweden, Report No. 1, 77 pp.
Rapp, A. (1987). Reflections on desertification 1977–1987: problems and prospects, *Desertification Control Bulletin*, **15**, 27–33.
Rapp, A., Le Houérou, H. N. and Lundholm, B. (1976). Can desert encroachment be stopped?, Ecological Bulletin 24, Swedish National Science Research Council, 241 pp.
Rasmusson, E. M. (1987). Global climate change and variability: effects on drought and desertification in Africa, in *Drought and Hunger in Africa* (Ed. M. H. Glantz), pp. 3–22, Cambridge University Press, Cambridge.
Renner, G. T. (1926). A famine zone in Africa: the Sudan, *Geographical Review*, **16**, 583–96.
Sabadell, J. E., Risley, E. M., Jorgenson, H. T. and Thornton, B. S. (1982). *Desertification in the United States: Status and Issues*, Bureau of Land Management, Department of the Interior, 277 pp.
Sauer, C. O. (1938). Destructive exploitation in modern colonial expansion, *International Geographical Congress, Amsterdam*, **III**, sec. IIIc, 494–9.
Schonken, J. D. (1924). Desiccation and how to measure it, *South African Journal of Science*, **21**, 131–48.
Schwarz, E. H. L. (1923). *The Kalahari or Thirstland Redemption*, Masker Miller, Cape Town and Oxford.
Sheridan, D. (1981). *Desertification of the United States*, p. 142, Council on Environmental Quality, Washington D.C.
Shiva, V. and Bandyopadhyay, J. (1986). Desertification in India: trends and counter-trends, *Desertification Control Bulletin*, **13**, 29–33.
Spooner, B. and Mann, H. S. (Eds.) (1982). *Desertification and Development, Dryland Ecology in Social Perspective*, Academic Press, London.
Stamp, L. D. (1940). The southern margin of the Sahara; comments on some recent studies on the question of desiccation in West Africa, *Geographical Review*, **30**, 297–300.

Stebbing, E. P. (1935). The encroaching Sahara: the threat to the West African colonies, *Geographical Journal*, **85**, 506–24.

Stein, A. (1938). Desiccation in Asia: a geographical question in the light of history, *Hungarian Quarterly*, **1938**, 13.

Stiles, D. N. (1983). Camel pastoralism and desertification in Northern Kenya, *Desertification Control Bulletin*, **8**, 2–8.

Stiles, D. (1984). Desertification: a question of linkage, *Desertification Control Bulletin*, **11**, 1–6.

Thompson, W. R. (1936). *Moisture and Farming in South Africa*, Central News Agency, 260 pp.

Tilho, J. (1928). Variations et desparition possible du Tchad, *Annales de Géographie*, **37**, 238–60.

Tyson, P. D. (1986). *Climatic Change and Variability in Southern Africa*, p. 220, Oxford University Press, Cape Town.

UNESCO/FAO (1977). *Desertification—Its Causes and Consequences*, Pergamon, Oxford.

Urvoy, Y. (1935). Terrasses et changements de climat quaternaires a l'est du Niger, *Annales de Géographie*, **44**, 254–63.

Van Reenen, R. J. (1923). A resumé of the drought problem in the Union of South Africa, *South African Journal of Science*, **20**, 178–92.

Van Ypersele, J. P. and Verstraete, M. M. (Eds.) (1986). Climate and desertification, *Climatic Change*, **9** (1/2), 1–258.

Verstraete, M. M. (1986). Defining desertification: a review, *Climatic Change*, **9**, 5–18.

Wadia, D. N. (1960). The post-glacial desiccation of Central Asia, National Institute of Sciences of India, Monograph, 26 pp.

Walker, J. and Rowntree, P. R. (1977). The effect of soil moisture on circulation and rainfall in a tropical model, *Quarterly Journal of the Royal Meteorological Society*, **103**, 29–46.

Walls, J. (1980). *Land, Man and Sand*, p. 336, Macmillan, New York.

Walsh, R. P. D., Hulme, M. and Campbell, M. D. (1988). Recent rainfall changes and their impact on hydrology and water supply in the semi-arid zone of the Sudan, *Geographical Journal*, **154**, 181–98.

Ward, M. N., Brooks, S. and Golland, C. K. (1988). Predictability of seasonal rainfall in the northern Nordeste region of Brazil, in *Recent Climatic Change* (Ed. S. Gregory), 237–51, Belhaven Press, London.

Warren, A. and Maizels, J. K. (1976). *Ecological Change and Desertification*, University College, London.

Wehmeier, E. (1980). Desertification processes and groundwater utilization in the northern Nefzaoua, Tunisia, *Stuttgarter Geographische Studien*, **95**, 125–43.

Wellington, J. H. (1967). *South West Africa and the Human Issues*, Clarendon Press, Oxford.

Wilson, J. F. (1865). Water supply in the basin of the River Orange or 'Gariep, South Africa, *Proceedings, Royal Geographical Society*, **35**, 106–12.

Worster, D. (1979). *Dust Bowl: The Southern Plains in the 1930s*, p. 277, Oxford University Press, New York.

Worthington, E. B. (1958). *Science in the Development of Africa*, p. 462, CCTA and CSA.

CHAPTER 2

The Control of Blowing Sand and Mobile Desert Dunes

A. WATSON

INTRODUCTION

The action of wind upon unconsolidated surficial materials can give rise to severe engineering problems in areas where there is insufficient vegetation to afford protection. The entrainment of clay, silt, and sand-sized particles by wind results in erosion of the land surface and the accumulation of sediment elsewhere. Moreover, particle movement can reduce visibility, can cause abrasion of natural and man-made structures, and may bring about the wholesale movement of dunes.

While many of these problems have been tackled in coastal environments in temperate regions (Brown, 1948; Hawk and Sharp, 1967; Savage and Woodhouse, 1968; Monohar and Bruun, 1970; Adriani and Terwindt, 1974; Davis, 1975; Knutson, 1977, 1980; Willetts and Phillips, 1978; Garés et al., 1979, 1980), comparatively little work has been undertaken in the arid zone. There are clear differences between these environments. In desert areas where fine-grained materials mantle the surface, aeolian erosion and sedimentation are difficult to combat because aridity restricts the potential vegetative cover. In contrast, in coastal areas the climate is usually conducive to vegetative stabilization provided suitable species are selected; indeed, in these areas sea water may be used to irrigate the plants (Ahmad, 1986). Viewed in the long term, however, some distinct similarities are apparent. The sand-flow regimes and dune systems are dynamic. In many deserts, the scale of the problems caused by moving sand precludes long-term solutions. The sand flow must be regarded as inexorable, since the source areas are too large or too distant to be stabilized. Similarly, in many coastal environments, especially those with barrier-island dune systems, one factor precludes any long-term solution to the problem of dune erosion—marine erosion of the

Techniques for Desert Reclamation
Edited by A. S. Goudie
© 1990 John Wiley & Sons Ltd.

dunes is unavoidable since it is a natural response to rising sea-level (Fink and Smith, 1974; Garés *et al.*, 1979; Nordstrom and Psuty, 1980). In effect, many of the sand control techniques discussed here must be viewed as necessarily short-term solutions, temporary holding exercises in the face of irresistible natural forces.

In most cases, the techniques for controlling blowing sand and moving dunes are well established. However, implementation of these procedures under field conditions frequently proves difficult. One reason for this is that empirical data on the rate of sand and dune movement are often scant (Jones *et al.*, 1986). Another is that 'real world' environmental phenomena— whether topography, aeolian dynamics, or the physical characteristics of the sand—are far more variable than those assumed for theoretical or simulated estimations of sand movement. Hence, any assessment of the effectiveness of these different sand control techniques must be based on an evaluation of their performance in the field as well as under controlled conditions. For these reasons, this discussion of the methods of sand control in deserts will concentrate on empirical studies.

One area that has seen a great deal of research into sand control and the practical application of many of the techniques is the Eastern Province of Saudi Arabia (Figure 2.1). The pioneering work there by Kerr and Nigra (1952) remains the most comprehensive in the field. While many of the principles underlying sand control in deserts have not changed since their work was undertaken, there is a need to review the field in the light of wider experience and recent technological advances. In the Eastern Province, the oilfields have spurred rapid economic development over the last thirty years. In addition to the petroleum installations, there are several large industrial complexes and numerous residential developments as well as the essential transport and communications facilities required to maintain the infrastructure. All of this is located within the Jafurah sand sea, a region of strong unidirectional winds, abundant sand dunes, and less than 80 mm of rainfall annually (Schyfsma, 1978). The sand control problems are immense; Fryberger *et al.* (1983, 1984) reported annual sand drift rates reaching 30 m³/m width.

The severity of the sand control problem in this area has prompted the adoption of a variety of procedures at individual locations. While the implementation of a range of techniques constituting a structured sand control policy is commendable, the approach can be self-defeating if the different philosophies of sand control are not appreciated. The aim of this discussion is to outline the fundamentals of aeolian sand control. These will be subdivided into those methods for controlling drifting sand and those for dealing with mobile dunes. Initially, some general comments are pertinent on the basic criteria for locating and designing facilities in areas of high sand flux.

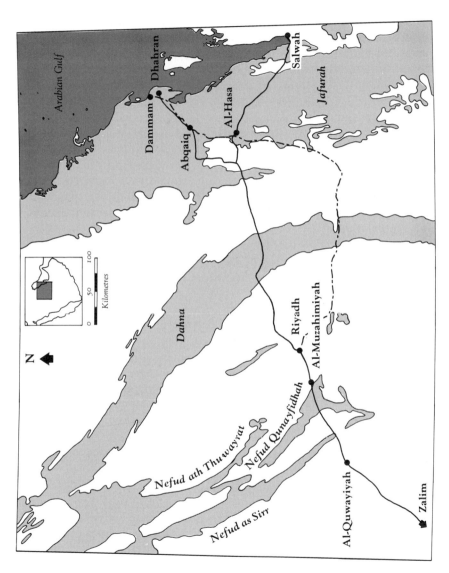

Figure 2.1 Eastern and central Saudi Arabia showing major sand bodies (lightly shaded), roads (solid lines), and railways (broken line)

DESIGN CRITERIA

One of the fundamental premises upon which the design of installations, whether roads, railways, individual buildings, or whole complexes, must be based is that the complete arrest of sand flow is not feasible. Hence, the design of most facilities must allow a degree of through-flow of sand. By following several basic principles, however, the potential for sand accumulation can be minimized. First, linear installations, such as pipelines and roads, should wherever possible run parallel to the direction of dominant sand drift. Clearly, this is not always possible, especially over large distances and in areas where sand drift is not unidirectional. Second, the facilities should be located in areas where the dominant sand-moving winds blow over gently rising ground. The increased ground-level wind strength on the windward side of a hill will promote sand transport rather than deposition. Third, the presence of features that disrupt the airflow should be minimized since they will stimulate sand deposition. Where such obstructions are unavoidable, they should be aerodynamically streamlined in order to prevent excessive sand accumulation.

Notwithstanding these preventative measures, there will always be instances when the free movement of sand cannot be tolerated or when additional procedures are required in the face of a severe problem.

DRIFTING SAND

Sand moving by surface creep or by saltation/reptation (Anderson, 1987) causes problems when the movement is hindered by a reduction in wind strength or by changes in the texture of the surface over which the sand is migrating. Four main approaches have been employed to avert these problems:

(a) promotion of the deposition of drifting sand (upwind of the problem area);
(b) enhancement of the transportation of sand (within the problem area);
(c) reduction of the sand supply; and
(d) deflection of the moving sand.

The procedures which have been employed to achieve these ends are numerous and varied. In some cases, the same techniques have been adopted to achieve apparently different objectives.

Promotion of the Deposition of Drifting Sand

This involves reducing the sediment load of the wind by decreasing its sand-transporting capacity. This can be undertaken in several different ways.

Ditches The excavation of pits and trenches upwind of an installation can afford a high degree of temporary protection from wind-borne sand. While any ditch will trap all sand grains moving as surface creep, to be fully effective it must be wider than the maximum horizontal jump of saltating grains. This may reach 3.0 to 4.0 m. The ditch must be sufficiently deep to prevent aeolian scouring of sand from the floor. Since the accumulation of sand will reduce the depth of the trench, it must be cleared of sand when its minimum effective depth is reached, or new ditches must be excavated. Moreover, the excavated material must be removed to a location where it will not present a hazard. In some instances, the excavation of borrow pits for aggregate or ballast can be undertaken upwind of the construction area, or in the path of dunes, thereby trapping some sand. In general, however, ditches are useful only for short-term protection.

Barriers and fences The construction of fences upwind of areas requiring protection from drifting sand has proved very successful. While the size, porosity, and morphology of sand fences are diverse, all types operate on the same principle. The fence creates a barrier to the wind and produces areas of reduced sand-carrying capacity ahead of and behind it. Ahead of a solid barrier, sand deposition occurs in a zone at a distance upwind from 0.4 to 2.0 times the height of the wall (Tsoar, 1983). With porous fences, however, most sand accumulation occurs in the lee—mainly in a zone four times as wide as the fence is high (Manohar and Bruun, 1970). By reducing the sediment load of the airstream, sand mobilization downwind is also reduced since there are fewer saltating grains to stimulate movement of surface particles. The porosity of the fence, that is the percentage of voids to total area, and the arrangement of the spaces affect the percentage of sand that is trapped (Phillips and Willetts, 1979). However, individual fence designs have different sand-trapping abilities at different wind velocities (Savage and Woodhouse, 1968; Manohar and Bruun, 1970). For this reason, fence design must be tailored to local environmental conditions. The alignment of the fences relative to the direction of resultant sand drift, and the number and spacing of fence rows are also critical (Kerr and Nigra, 1952; Willetts and Phillips, 1978). Snyder and Pinet (1980) found that in coastal environments the broad, low dunes created by zig-zag fences were preferable to the higher dunes formed by straight fences since they are less prone to subsequent erosion. Moreover, high dunes, especially those with a concave slope profile near the crest, are less stable than flatter, more convex dunes (Lai and Wu, 1978) (Figure 2.2). In areas where the sand-bearing winds are not unidirectional, fence arrangements must be designed accordingly. Davis (1975), for example, advocated diamond-shaped fence enclosures along the Florida coast.

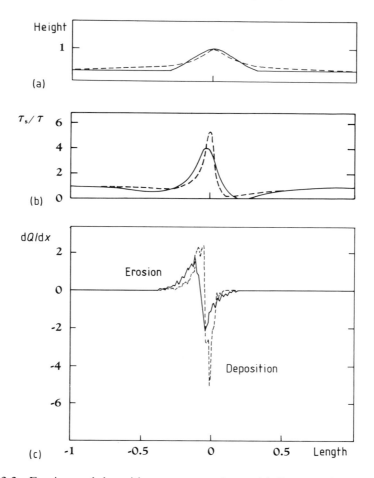

Figure 2.2   Erosion and deposition across two dunes. (a) Dune profiles—solid line shows a dune with a mainly convex profile and dashed line shows a dune with a mainly concave profile. (b) Variation in shear stress—τ_s is shear stress over a sloping surface, τ is shear stress over a flat surface when shear velocity is 0.50 m/s on the flat surface upwind of the dune. (c) Variations in calculated erosion and deposition (dQ/dx) across the profiles shown in (a) when shear velocity is 0.50 m/s and the sand grain diameter is 0.25 mm (after Lai and Wu, 1978)

The most common type of fencing employed along coasts, and also in many desert areas, is the slat-type comprising 100 mm wide vertical slats set 100 mm apart—giving a porosity of 50%. It has been shown that wooden-slat fences are more effective at trapping sand than less-rigid fences made of fabric (Figure 2.3), even when fence porosity is the same (Savage and Woodhouse, 1968). In trials on coastal dunes, Knutson (1980) found that over a seven-year

Figure 2.3 Fabric sand fencing with horizontal strips giving a 50% porosity. The fence is erected on a sabkha in the Jafurah sand sea in eastern Saudi Arabia

period sand fencing was slightly more effective at trapping sand than were vegetation plots. The fences, however, required regular maintenance and periodic replacement as they became buried by accumulating sand.

By reducing the wind's sand-carrying capacity in the lee of fences, the potential to transport sand is diminished. Manohar and Bruun (1970) found that the percentage of sand trapped by rows of 40 mm wide slat fence varied from 60% when wind speeds were 10 m/s, to 16% when speeds were 18 m/s. In areas experiencing a wide range of wind speeds, the optimum fence porosity is 40% (Savage and Woodhouse, 1968). However, with wind speed greater than 18 m/s no sand is trapped by a single fence of 40% porosity. Under these conditions, a double fence will trap about 30% of the sand in motion (Manohar and Bruun, 1970). Multiple fences can trap more than 80% of wind-borne sand, even under variable wind conditions, if optimal designs are employed. On the basis of the drift rates determined by Fryberger *et al.* (1983), about 24 m^3 of sand could collect annually behind each linear metre of fence in the Eastern Province of Saudi Arabia. Hence, the fences must be located in areas where the creation of a large artificial dune will not pose problems. Clearly, this depends upon the time over which sand control is required and upon the nature of the installations being protected.

Figure 2.4   Sand accumulation behind palm-leaf fences on a low dune in the Jafurah sand sea. Sand drifting from left to right has buried the first three rows of fences

Since the fences will eventually be buried by the trapped sand (Figure 2.4), new fences must be erected on the accumulating mound. The utilization of fences over a long period constitutes a commitment to a sand control policy based upon dune building (Figure 2.5). Very large dunes may be created which themselves need careful management to prevent encroachment on the installations downwind (Kerr and Nigra, 1952; Trossel, 1981).

Vegetation belts   By establishing belts of trees to act as barriers to drifting sand, the frequent renewal of sand fences is obviated (Davis, 1975). It is essential, however, to select trees species which have growth rates greater than the rate of sand accumulation and which also have a bushy shape. In warm semi-arid and arid environments, *Tamarix* and *Eucalyptus* species have proven most successful. In areas where the tree belts are of limited extent, irrigated plantations are preferable since both growth rates and survival rates are high. This is undertaken along major roads in Abu Dhabi in the United Arab Emirates. In Saudi Arabia, however, a programme of non-irrigated cultivation of *Tamarix aphylla* has been implemented. Notable success has

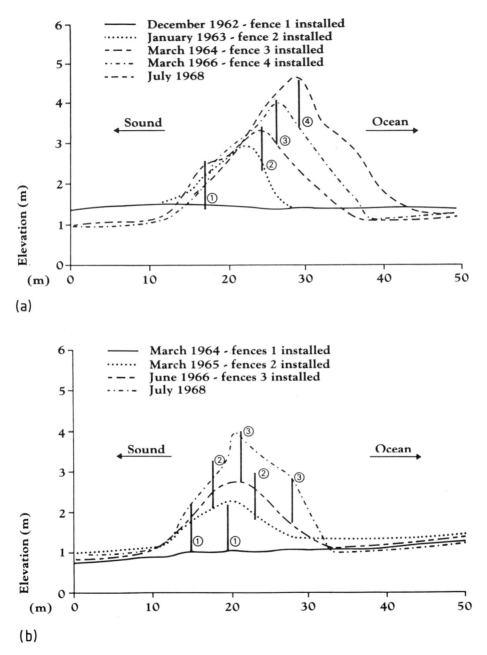

Figure 2.5   Dune building using sand fences on the coast of North Carolina, USA.
(a) Using single fences; (b) using double fences (after Savage and Woodhouse, 1968)

Figure 2.6   Belt of non-irrigated *Tamarix* trees north of the Al-Hasa Oasis, Saudi Arabia. Neither rapid sand accumulation nor dune encroachment have hindered the trees' growth

been achieved in the Al-Hasa Oasis (Stevens, 1974; Hidore and Albokhair, 1982) (Figures 2.1 and 2.6).

There is at present little long-term data on the viability of non-irrigated vegetative sand control schemes. Again, the selection of appropriate species is critical in order to minimize the risk of the complete destruction of the plantations as a result of disease, pests, or aberrant weather conditions (Brown and Hafenrichter, 1962). There is also a question mark over the long-term effectiveness of non-irrigated tree belts. Gupta (1979) showed that the soil moisture in a sand dune is severely depleted in the presence of vegetation. Since a mature tree will require more water, its root system will search out groundwater. In areas peripheral to saline sabkhas, the groundwater may be sufficiently salty to kill the tree. This may not occur for several years, a period during which the tree belt will have trapped sand. The death of the trees could then lead to the mobilization of the sand mound.

Notwithstanding these potential problems, vegetative sand control techniques have been widely implemented because they are often relatively inexpensive and are less disruptive of the environment than other methods of stabilization. In China, along the Bagoton–Lanchou railway where it crosses the Tengeri sand sea, 300 to 500 m wide vegetated strips on the upwind side of the tracks trap most of the wind-blown sand (Liu Shu, 1986).

Enhancement of the Transportation of Sand

This approach to sand control is in many ways incompatible with the afore-mentioned approach of sand entrapment. The principal objective is to increase the wind's sand-moving capacity by either increasing wind velocity or reducing the ability of the surface to trap moving grains.

Aerodynamic streamlining   By shaping the land surface either to increase the wind velocity over the surface or to remove areas of reduced sand-carrying capacity, sand deposition can be greatly reduced. While specific designs may be developed for individual locations, generalized optimal profiles for embankments and cuttings have been established. To prevent sand accumu-lation on an embanked road or railway, the windward slope should be graded no steeper than 1:6 (22%). This corresponds roughly to the windward slope of a barchan dune. Embankment slopes of up to 1:3½ (about 37%) have been proposed (Ove Arup and Partners, no date) but their effectiveness in areas with high sand-transport rates has not been tested. Such an embank-ment will act as a sand trap, causing accumulation of sand on the windward side in the same way as a solid barrier. In time, this deposit would probably require surface stabilization to prevent scouring and the entrainment of sand. Moreover, such an embankment would probably require crash barriers adja-cent to the pavement. These would promote sand accumulation on the road surface, as would separation in the airflow above the pacement as a result of the steep inclination of the windward slope. It should be stressed that the break in slope at the top of any embankment should be rounded not angular. Any abrupt changes in slope promote separation in the airflow creating a bubble of reduced wind strength wherein sand is deposited.

The upwind slopes of road and rail cuttings must be inclined at shallower angles than the windward slopes of embankments if sand accumulation is to be averted. While wind speeds will increase over an upward-sloping surface, increasing the sand-carrying potential, this potential is lost over a downward incline. Slopes not exceeding 1:10 (13%) have been advocated (BMMK and Partners, no date; Ove Arup and Partners, no date).

Surface treatment   Mantling the land surface to enhance surface creep and sand grain saltation is widely used to prevent sand accumulation. The procedure has the secondary effect of protecting the surface from erosion, thereby reducing the potential sand supply further downwind.

There are numerous different techniques of surface treatment (Aripov and Nuryev, 1982) and, as is described below, their objectives are frequently very different. For example, it is possible to create a rough surface texture, using coarse aggregate, which acts as a sand trap. Similarly, the spraying of waxy oil on the surface can cause sand to adhere there. These techniques are rarely used to promote sand deposition since the sand-storage capacity per unit area is very small compared with that for ditches or fences. More common are surface treatment techniques which create a smooth, resilient surface. This reduces the number of crevices in which sand grains can lodge and creates a surface which does not absorb the kinetic energy of the saltating grains. Since the energy is not transmitted to other sand grains on the surface, the grains' movement is not hindered. Hence sand will move across roads, runways, and similar non-obstructive features, rather than accumulate upon them.

A variety of materials have been used in desert areas including asphalt, synthetic latex, polyvinyl, sodium silicate, and gelatine. Most have short effective life spans, generally about one to five years. The longevity of any chemical treatment increases with the resilience and thickness of the material. In most cases, materials that harden to form an erosion-resistant surface suffer from low surface penetration and, therefore, are prone to undercutting as a result of scour. In order to achieve better penetration, chemicals can be injected into the sand rather than poured or sprayed onto the surface. Pressure injection techniques require special equipment, increasing the application costs. A substance that can be sprayed onto the surface and still achieve penetration and produce a hard, resistant surface would be ideal. To this end emulsified resins are attracting increasing attention. In Central Asia, recycled motor oil and sulphate–alcohol distillery residues which have been emulsified with polyvinyl acetate are used as sand stabilizers (Arnagel'dyev and Kurbanov, 1984). Some substances can be treated so as not to discolour the land surface while others have been employed in conjunction with hydroseeding schemes.

Natural oils tend to oxidize and lose their effectiveness as binding agents but when emulsified, especially when diluted with water, they can be inexpensive and can be applied on irregular surfaces by spraying. Asphalts have been used by ARAMCO and the Ministry of Communications in Saudi Arabia and are used widely in Central Asia along roads and pipelines (Aripov and Nuryev, 1982; Petrov, 1983; Arnagel'dyev and Kurbanov, 1984). However, they present several disadvantages: their protective action is limited, the sticky remnants are found even tens of years after the asphalt was applied, and the effect on the environment is unpleasant and unsightly.

Panelling   The erection of panels which channel airflow over a road surface or around an installation has been undertaken by ARAMCO in Saudi Arabia. It is employed only where immediate protection from blowing sand is necessary and where the rate of sand deposition is low (Trossel, 1981). The method is expensive since either the panelling needs replacing as it is buried by the accumulating sand or the sand must be removed mechanically (Clements *et al.*, 1963). While panels can be arranged to direct the airflow over a surface, keeping it sand free, the local increase in wind velocity implies a reduction in wind speed elsewhere. Sand deposition will occur in the zone of reduced wind speed and consideration must be given to the consequences of this when the technique is employed.

Reduction of the Sand Supply

In areas where intense sand movement occurs, enhancing sand deposition or transport may be impractical or impossible. The rate of accumulation behind fences may be so great that they must be frequently replaced or the movement of sand across a road may severely hinder visibility. In such instances, it may prove preferable to identify the main sources of sand and stabilize the deflating sand body. The three types of procedure outlined here encompass the main methods of protecting unconsolidated sediments from aeolian erosion (for a detailed discussion see Chepil and Woodruff, 1963).

Surface treatment   Whereas the various forms of surface treatment described above are intended to aid sand transport over a surface, the following are intended to do little more than protect the underlying unconsolidated materials. A variety of techniques have been employed including the following:

(a) Increasing the grain size of the surface material: In many desert areas, the sand surface is protected from deflation by a natural lag of coarse-grained material. This layer develops as finer particles are deflated or when the fines are eluviated (McFadden *et al.*, 1987) leaving a gravelly residue at the surface. The armoured surface is termed a desert pavement. If gravel-sized particles are spread over a surface, they create a protective blanket, preventing deflation. In order to protect the surface in the strongest winds, a minimum grain diameter of about 4.0 mm is required since smaller particles will be affected by creep. Lai and Wu (1978) noted that an increase in the size of sand grains only slightly decreases sand transport when wind velocities are high. The artificial lag gravel should consist of materials which are not susceptible to weathering or abrasion. The layer should be no less than 50 mm thick since if it is disrupted, rapid scouring of the underlying material will occur. Moreover, if the cover of coarse particles is sparse, it adds roughness to the surface thereby increasing turbulence in the airflow and enhancing the potential for sand entrainment (Lai and Wu, 1978; Logie, 1981). A more closely spaced cover of the same coarse material may prove an effective blanket preventing sand mobilization (Logie, 1981). Similar protective mantles may be composed of grains with a flatter, more platy shape or of greater specific gravity than the underlying sand (Willets, 1983).

It has been noted that clayey soils often require higher wind velocities in order to induce their erosion (Chepil and Woodruff, 1963; Gillette *et al.*, 1982). This effect results from the cohesion between clay particles, following wetting, which creates an encrusted surface. The threshold velocity (the shear velocity at which particle movement commences) may be increased from about 0.60 to 1.9 m/s by crusting (Gillette *et al.*, 1980). Generally, the higher the clay content and the higher the exchangeable sodium percentage, the greater the crustal strength. However, should the protective crust be ruptured, the underlying material may be highly susceptible to erosion owing to the small grain size (Gillette and Walker, 1977).

(b) Chemical treatment of the surface: A wide variety of chemicals have been used as sand stabilizers. They either form a protective coating over the surface or create cohesion between the surface particles. Cohesion is achieved through the formation of an adhesive film between the sand grains. The different binding mechanisms are not in themselves significant from a practical point of view. The critical factor is the life expectancy of the treatment, especially when this is related to the cost of materials and application. Additional factors such as the visual appearance of the treated area, the effect the chemicals have on plant life (Polyakova, 1976), and their susceptibility to leaching or erosion by rainwater (Gabriels *et al.*, 1974) must also be taken into account.

The use of chemical sand stabilizers is becoming increasingly widespread in desert regions. In Saudi Arabia, one company treated nearly 35 km^2 over the six years from 1978 to 1983; about 50 km^2 have been treated in Central Asia (Petrov, 1983). The spray application of plant seeds admixed with sand stabilizers combined with fertilizers (Polyakova, 1976), humidifying substances, and seed hormone activators has also been attempted. To date such schemes have been restricted to areas where the plants can be irrigated once they have germinated.

(c) Application of a surface layer of oil or asphalt: The main advantage of the application of crude oils and asphalts over that of chemicals is one of cost. In the Eastern Province of Saudi Arabia, suitable asphalts and

Figure 2.7   Sand surface stabilization using cutback asphalt around an oil facility

cutbacks are available at wellheads and refineries (Figure 2.7); in Central Asia, they are obtained from the Mangyshlak and Dzharkurgan oilfields of Uzbekistan (Mirkhmedov, 1983). Moreover, the variety of crude oils provides a choice of substances. Trossel (1981) reported the use of low gravity asphaltic oils and high gravity waxy oil by ARAMCO. The latter is preferable because it achieves deep penetration and does not form a brittle crust which is susceptible to deflation. However, the waxy oil's property of remaining sticky and flexible makes it unsuitable in areas where personnel and vehicles must traverse the treated surface.

Local conditions are critical in the selection of a suitable sand stabilizer. While in some localities it may be feasible to treat the surface at yearly intervals, in remote places the initial treatment may be required to last ten years or more. Oils and asphalts are prone to oxidation over a period of one to two years and once this has occurred the treatment rapidly becomes ineffective. Cutback asphalts are less prone to oxidation and with additives, such as liquid latex, their tendency to become brittle is reduced. Despite these improvements, the performance of cutbacks is limited by their low surface penetration. Other materials are prone to disintegrate when they are bombarded by sand grains (Azizov and

Atabaev, 1987). In Saudi Arabia, extensive trials of cationic emulsified asphalts have given favourable results but a major drawback with all of these oils and asphalts is that they are toxic to many plants. Nevertheless, in Central Asia, heavy crude oils and emulsified asphalts have been used in conjunction with vegetative sand stabilization techniques (Mirakhmedov, 1983; Nuryev *et al.*, 1985).

(d) Binding the surface using natural materials: Several other types of surface treatment have been employed to prevent deflation. By increasing the moisture content of sand, its erodibility can be lowered significantly (Bisal and Hsieh, 1966). Svasek and Terwindt (1974) showed that while dry sand is moved by the wind when the shear velocity is greater than 0.10 m/s, sand containing about 0.80% water (by volume) does not move until the shear velocity is about 0.65 m/s. The difference suggests that the wind speed at about 2.0 m above the sand surface must reach 5.0 m/s in order to move dry sand, while it must reach about 15 to 20 m/s to move damp sand. In contrast, Sarre (1988) found that moisture levels of up to 14% had no discernible effect on sand transport rates on a beach. Bisal and Hsieh (1966) pointed out that the quantities of water required to prevent deflation of different soils were larger when the materials contained abundant fine particles. At wind speeds of 12.5 m/s the moisture content needed to prevent erosion was 7.9% for sandy loam, 10.8% for loam, and 15.7% for clay soils. It is interesting to note that Lyles and Schrandt (1972) reported that rainfall duration had little significant effect on the erodibility of soil by wind. Moreover, Hidore and Albokhair (1982) found that in Al-Hasa (Figure 2.1), sand drift rates were reduced for only about 24 hours following rainfall. Continual application of large quantities of water would be required to affect stabilization of a sand body in eastern Saudi Arabia, especially during the summer months when sand surface temperatures reach 70°C and strong northerly winds prevail (Trossel, 1981).

The formation of a surface salt crust, salcrete, can protect an unconsolidated sand body from deflation at wind speeds of up to 12 m/s (Pye, 1980). Such crusts can be created artificially by spraying saline water over the sand surface. Crystallization of the salts as the water evaporates binds the sand grains (Lyles and Schrandt, 1972; Gillette *et al.*, 1980, 1982), but the formation of efflorescences may increase surface roughness and consequently lower the threshold velocity of the surface materials (Nickling and Ecclestone, 1981; Nickling, 1984). Moreover, local breaching of the crust may initiate rapid scouring of the underlying sand. These salt crusts are susceptible to dissolution by rainwater and can disintegrate under hot, dry conditions if readily dehydrated salts, such as calcium and sodium sulphates, predominate.

In the Eastern Province of Saudi Arabia, there are large quantities of saline water available. The water table is usually just 1.0 to 2.0 m below the surface of the numerous sabkhas which dot the coastal plain (Johnson *et al.*, 1978; Fryberger *et al.*, 1983). While this water is of no use to industry or agriculture, its exploitation could prove detrimental to the cause of sand control. This is because most of the sabkhas are deflationary depressions, the present surfaces resisting erosion through the combined effect of a surface salt crust and a high moisture content (Fryberger *et al.*, 1988). Should the water table be lowered, large quantities of sand would be deflated once the surficial crust is dissolved and the salts leached from the surface.

Fences   The use of fences to protect a source area from deflation is essentially the same as their use to trap sand as is described above. The practical number of fence rows in a multiple-fence belt is about four. If more fences are installed, the added trapping efficiency is negligible (Figure 2.4). Ideally, the fence height should be determined on the evidence of annual rates of sand movement. Since optimum spacing of fence rows is four times the height of the fences (Manohar and Bruun, 1970), 1.0 m high fences have a maximum capacity to trap about 16 m³/m width—assuming four rows of fence each 4.0 m apart. Once the fenced area has been buried by sand, the next fences installed on top of the sand mound will have a greater sand-trapping capacity since the volume impounded by the barriers varies as the square of the height of the barrier. The second installation of fences would create a barrier twice the original height, thereby increasing the life expectancy of the second fence system to four times that of the first (Trossel, 1981). This is upheld by the empirical studies of Savage and Woodhouse (1968) along the coast of North Carolina (Figure 2.5) and Kerr and Nigra (1952) around Dhahran (Figure 2.1).

Vegetation   The introduction of vegetation to protect areas from deflation or to trap sand moving away from source areas can take two forms. First, belts of trees can be established to act as self-renewing fence systems, as described above. Second, a ground cover can be established in the area experiencing deflation to protect the surface from erosion; for example, Prishchepa (1984) noted that deflation decreased around the Karakum Canal in Central Asia when rising ground-water levels led to vegetation colonizing formerly bare sand surfaces. Much work on vegetation stabilization has been undertaken on the coastal dunes of the Netherlands (Adriani and Terwindt, 1974) and the eastern United States (USDA, no date, 1977, 1982).

In comparison with fences, vegetation plots are less efficient at trapping sand (Knutson, 1980), though in part this is because an initial period is required before the plants become established. The experience in coastal areas has shown that once grass plots are established they are far easier to maintain than fences. Indeed, plots of American beachgrass (*Ammophila breviligulata*) can trap over 1.25 m of sand annually (Jagschitz and Wakefield, 1971) and will often extend the cover beyond the zone of planting (Knutson, 1977, 1980). Savage and Woodhouse (1968) found that 11 to 13 m wide belts of American beachgrass not only protected coastal dunes from deflation but also trapped all the sand drifting across the belt. Grass growth was sufficient to prevent burial of the plants by sand. Similar success has been reported from the North Sea coast of Scotland (Ritchie, 1974). However, in desert environments the threat of extended periods of drought and also the irregular rate of sand flow may limit the effectiveness of such techniques. In Saudi Arabia, for example, Hidore and Albokhair (1982) noted that over a monitoring period of several months, 25% of all sand movement occurred during one storm lasting only 43 hours; under such conditions, rapid sand accumulation may bury the plants, precluding their survival. Also, in many deserts, vegetative sand control schemes require additional measures to protect the plots from grazing animals, particularly camels, goats, and sheep (Liu Shu, 1986). Hence, these techniques require continual monitoring and maintenance; they should not be considered 'a method with (an) unlimited lifespan', as Tsoar and Zohar (1985) suggested.

Another drawback with this approach to sand stabilization is the difficulty in finding plant species capable of withstanding severe drought. High rates of sand accumulation or deflation, excessive grazing, high soil moisture salinities, and a shortage of irrigation water can limit the value of the technique. Suitable indigenous grass and sedge species are present in the Eastern Province of Saudi Arabia but lengthy field testing is required before their relative merits can be assessed. The principal species are *Stipagrostis plumosa*, *Cyperus conglumeratus*, *Panicum turgidum*, and *Rhanterium eppoposum*.

Deflection of the Moving Sand

The final approach to solving the problem of drifting sand involves employing fences, barriers, and tree belts to deflect the sandstream. ARAMCO has used the procedure in the Eastern Province to protect isolated buildings and wellheads. It is not advocated over large areas or along roads and pipelines (Trossel, 1981).

Fences and barriers   Two types of fence alignment have been employed. The first consists of a fence slanted at about 45° to the direction of sand drift, the

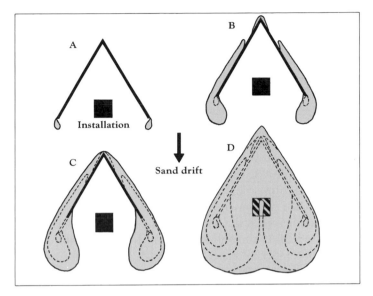

Figure 2.8   Performance of V-shaped diversion barriers (after Trossel, 1981)

second comprises a V-shaped barrier pointing into the sandstream (Figure 2.8). In both cases, the degree of protection and the effective life span depend on the porosity and height of the fences. High barriers of low porosity will divert the sand flow until the sand accumulation on the windward side overwhelms the barrier. Multiple fence rows are more effective than single rows, especially at high wind speeds (greater than 18 m/s). Such fences not only trap sand but also deflect the airstream, thereby removing some of the accumulated sand. However, any structure that promotes sand deposition will have a limited life span and, therefore, diversion fences must be regularly cleared of sand and are recommended only where temporary protection is required.

Tree belts   In some ways preferable to fences, tree belts which deflect sand from its dominant direction of movement have been employed along railways and roads in the USSR (Shirmamedov, 1978) and the USA (Clements *et al.*, 1963). While they can provide a high degree of protection, the accumulated sand may give rise to additional problems. Moreover, should the tree belt be destroyed as a result of drought, lack of irrigation, or pests, the artificial sand bank may become unstable. Blowouts may develop (Jungerius *et al.*, 1981) and related parabolic dunes can migrate downwind. The problems associated with mobile dunes constitute another major and distinct aspect of aeolian sand control.

DESERT DUNES

Though there are clear morphological similarities between sand ripples (which are the smallest aeolian bedform) and some dunes, they do not form an evolutionary continuum. While obstacle dunes may begin to develop as small-scale features, most dune types that exist as independent aerodynamic forms are several orders of magnitude larger than ripples. Distinct barchanoid dune forms are rarely less than 1.0 to 2.0 m high and some dune types reach heights of more than 300 m. In contrast with ripples that form spontaneously, dunes develop slowly and are in equilibrium with longer-term environmental conditions. Active dunes are by no means confined to the earth's warm desert regions. Sand dunes forming extensive ergs (sand seas) occur on Mars where strong winds mobilize particulate material despite the thin atmosphere and frozen land surface (Breed *et al.*, 1979; Tsoar *et al.*, 1979). Dunes resembling those which are typical in the sub-tropics are found in Antarctica (Lindsay, 1973) as well as in temperate zone deserts such as the Gobi—which lies at a higher latitude than Boston, Massachusetts. Moreover, relict dunes are widespread in the tropics (Talbot, 1984). In the past, particularly during the late Pleistocene, dune systems covered a considerably larger area of the globe than today (Thomas and Goudie, 1984). Coatsal dunes are in some ways distinct because their genesis is closely related to specific sand sources and vegetational environments.

In deserts, dunes develop as sand grains being transported by the wind are deposited and accumulate preferentially within a limited area. This initial accumulation may occur in a zone of reduced wind strength in a depression or in the lee of an obstacle (Hesp, 1981; Besler, 1984). Once the surface is covered by a layer of sand grains, more grains accumulate when they transfer their momentum to the surface grains rather than conserving their momentum by bouncing over the surrounding solid or pebbly surface. The reduction in wind velocity immediately above the sandy surface owing to sand movement further increases the potential for accumulation of incoming grains (Bagnold, 1941, pp. 57–9). Provided that more sand accretes through these processes of 'splashdown' and wind retardation than is deflated from the downwind margin, the sand patch will accrete (Mabbutt, 1977, p. 223). The patch of sand may be classified as a dune when it reaches certain minimum dimensions. Depending on such factors as wind velocity, variability in wind direction, rate of sand supply, and the nature of the terrain, the sand accumulations evolve into different dune forms (Hack, 1941) (Figure 2.9). A number of dune classifications have been proposed. For example, Bagnold (1941, p. 189) classified dunes into transverse or longitudinal types depending on whether their crests are oriented perpendicular or parallel to the dominant direction of the sand-moving wind. Rubin and Hunter (1985) classified dunes aligned at $90° ± 15°$ and $270° ± 15°$ to the direction of sand flow as transverse; those at $0° ± 15°$ and $180° ± 15°$ as longitudinal; and all others as oblique.

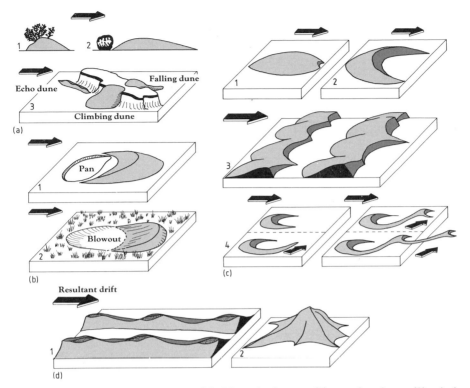

Figure 2.9   Selected dune forms. (a) Obstacle dunes—(1) coppice dune; (2) wind-shadow dune; (3) dune forms in the vicinity of a plateau. (b) Deflation forms (1) lunette; (2) parabolic dune. (c) Transverse dunes—(1) incipient barchan developing from a dome dune; (2) barchan dune; (3) transverse dune ridges; (4) formation of a longitudinal (seif) dune from a barchan under a bidirectional wind regime (after Lancaster, 1980). Tsoar (1984) held that the arm downwind of the oblique flow becomes elongated. (d) (1) Longitudinal dune ridges; (2) star dune

Mader and Yardley (1985) termed simple, migrating, transverse forms, *primary dunes*; longitudinal forms produced by morphological modification and migration, *secondary dunes*; and complex forms such as star dunes produced by merging of transverse and longitudinal elements—as well as modification and migration—*tertiary dunes*. For the purposes of this brief description, these forms are classified into four categories on the basis of their morphological and genetic characteristics: obstacle or topographic dunes; deflation forms; transverse dunes; and longitudinal dunes (Figure 2.9).

Obstacle and Topographic Dunes

The simplest type of sand accumulation is that associated with the zones of reduced wind strength ahead of or in the lee of bushes, boulders, or hills.

Reversal of the direction of airflow in the lee of such roughness elements results in the deposition of grains from a sand-laden airstream (Hesp, 1981). Coppice dunes, sometimes called elephant head dunes, are characteristically vegetated. The grasses or shrubs which colonize them trap wind-borne sand thereby perpetuating the landform. The coppice will often possess a sand plume or wind shadow dune which tails away from the mound in the downwind direction. On a larger scale, such wind shadow dunes occur in the lee of hills and plateaux where they are known as falling dunes. In the area of reduced wind speeds upwind of such topographic obstacles, sand accumulations form climbing dunes (Evans, 1962; Howard, 1985). Upwind of cliffs inclined at greater than about 55°, the development of a vortex keeps the face free of sand but an echo dune forms ahead of the obstruction. All these dune forms are static since their formation is dependent upon fixed obstacles.

Deflation Forms

In areas with high rates of sand supply owing to deflation of an area immediately upwind, parabolic dunes and lunettes can develop. The former result when a vegetated dune or sand sheet is locally devegetated; a blowout develops (Jungerius *et al.*, 1981) and large quantities of sand accumulate on its downwind margin. If a sufficient volume of sand collects, the parabolic dune may become mobile. This movement gives rise to the typical U-shape or V-shape of this dune form (Breed and Grow, 1979). The migrating mound of sand—the nose of the parabolic dune—may evolve into a transverse dune if the sand supply and wind regime are conducive. Fields of parabolic dunes are common in many sandy environments. Often, the degradation of the vegetation cover which prompts their development results from overgrazing or human interference (Anton, 1983; Khodzhaev, 1983).

Lunettes have similar morphologies to parabolic dunes. Generally, however, they are fixed since the sediment they are formed from is derived from deflating depressions or pans immediately upwind. Some lunettes, such as those flanking large sabkhas in central Tunisia, reach heights of 50 m or more.

Transverse Dunes

Dunes with crests aligned perpendicular to the wind differ from obstacle dunes and deflation forms in that they are not fixed; they migrate in the direction of the dominant wind. Indeed, they are usually confined to areas with unidirectional winds. The simplest form of mobile dune is the dome dune. Though strictly such circular or elliptical dunes are neither transverse or longitudinal, there is evidence that they may evolve into or from barchanoid dunes which are transverse forms (McKee, 1979a).

As an incipient dune develops from a sand patch, the airflow on the windward side is compressed and its velocity increases (Lancaster, 1985; Tsoar, 1985). As a result, the capacity of the wind to move sand up the slope also increases once the wind has responded to the additional drag resulting from sand transport. The positions of maximum erosion and deposition are not, however, solely dependent on either local surface slope as Bagnold (1941, pp. 198–201) suggested or wind velocity (Lancaster, 1985; Tsoar, 1985). Rather, the positions vary with surface shear stress (Lai and Wu, 1978) (Figure 2.2). Moreover, the retardation of flow velocity near the surface owing to sand transport (Ungar and Haff, 1987) further complicates calculation of the sand transport rate using wind speed data (Mulligan, 1988). Though Bagnold (1941, p. 119) predicted that the maximum rate of sand deposition would be in the zone of reduced wind strength beyond the dune's crest, where the airflow has separated from the surface, Lai and Wu (1978) found that the highest deposition rates can occur at the top of the dune. This will depend upon the shape of the windward flank and upon variations in shear velocity over the surface (Mercer and Haque, 1973). This mechanism of crestal accretion is significant since it can account for the vertical growth of dunes. If wind speed was the only factor influencing sand movement, sand at the crest would always be prone to deflation (Watson, 1987) because wind velocities are greatest at the dune's highest point (Lancaster, 1985).

At the crest of the dune, the change in surface inclination from an upward to a downward slope may promote detachment of the airflow. This separation of the flow from the surface in the lee of the crest is most likely to occur when there is a sharp break in slope. However, it can occur at rounded dune crests under conditions of high-velocity airflow. The establishment of a separation wake results in rapid deposition of sand in the lee of the crest.

Lai and Wu (1978) held that deposition at the crest steepens the windward slope, increasing shear stress, sand transport, and dune height. Under these conditions of positive feedback, sand cannot accumulate indefinitely; it is either blown from the crest or the leeward slope becomes oversteepened. At a critical angle—the angle of initial yield—which in the case of dry, sand-sized particles is about $34°$, the leeward slope becomes unstable and avalanching occurs on that side of the dune; a slipface is formed (Figure 2.10). After avalanching, this surface is inclined at the angle of residual shear, about $31°$ to $32°$; so the angle of repose of dry sand on a slipface lies between $31°$ and $34°$. If the angle of initial yield is not reached, because wind speeds are too low or because the sand mound is too small, a dome dune will persist. Some dome dunes, however, appear to be restricted to areas with very high wind speeds (Breed and Grow, 1979) or high rates of sand supply (McKee, 1979b). This suggests that oversteepening owing to sand deposition at the dune crest (Lai and Wu, 1978) may be as significant in slipface genesis as oversteepening

Figure 2.10   Slipface of a 35 m high barchan in the Jafurah sand sea, Saudi Arabia

resulting from flow separation and sand deposition downwind of the crest (Wilson, 1972). Once a slipface has developed, a dome dune has an incipient barchanoid morphology (Figure 2.11).

The slipface of a barchanoid dune acts as a sand trap. The sharp break in slope at the top of the slipface (the brink) maintains separation in the airflow and creates a bubble of lower wind speeds in the lee of the dune. Not only do creeping sand grains avalanche down the slipface, but airborne particles rain down into the pocket of comparatively still air (Hunter, 1985). It has been suggested (Whitney, 1978, 1983; Warren and Knott, 1983) that a large vortex or return flow in the lee of the brink helps perpetuate the steep angle of the slipface. The occurrence of small deflationary sabkhas in the lee of some barchans in eastern Saudi Arabia may be related to such eddying within the separation wake.

Barchan dunes advance through a process of erosion on the windward side and deposition on the leeward (the slipface) (Figure 2.10) (Howard *et al.*, 1978). Since

$$c = \frac{Q}{yh}$$

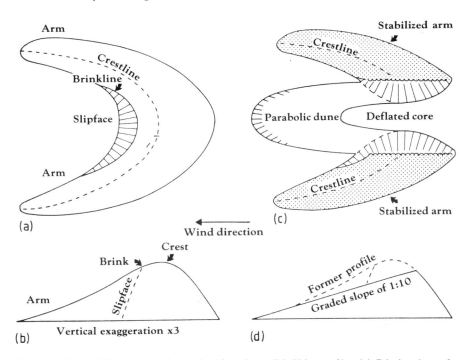

Figure 2.11   (a) Plan view of a typical barchan. (b) Side profile. (c) Dissipation of a barchan by stabilization of the arms (horns). (d) Side profile of a barchan that has been reshaped to minimize the sand-trapping effect of the slipface

where c = distance of advance in unit time
      Q = rate of sand flow at the brink
      y = specific weight of loosely packed, bulk sand
      h = height of the brink

'the rate of advance of the dune must vary directly as the rate of sand move-ment over the brink, and inversely as the height of the slipface' (Bagnold, 1941, p. 204). The volume of sand that must be deposited to maintain the dune's form is greatest at the highest portion of the dune, and initially this part advances slower than the sides—where the slipface is lower. By advancing more at the sides, the brinkline becomes curved and the character-istic horned shape of the barchan develops (Figures 2.11 and 2.12). However, the wind velocity increases with height on the dune and 'the rate (of sand removal and deposition) is a maximum where the surface angle is steepest' (Bagnold, 1941, p. 201). So, the steeper, higher portion of the dune, where rates of sand transport are greatest, will keep pace with the lower horns once equilibrium is achieved through a reduction in the inclination of the windward flanks of the horns.

Figure 2.12   Barchan dunes up to about 20 m high moving north to south (top to bottom) across a sabkha west of Dammam, Saudi Arabia (see Figure 2.1). This aerial photograph shows an area of about 2.0 by 2.0 km. Note the plumes of deflated sand extending from the horns of the barchans and the arcuate patterns in the wake of the dunes—these are produced by cementation of the basal portion of the slipface (foreset beds) by halite precipitated from evaporating sabkha water. Also note the development of incipient transverse ridges to the bottom left

Factors such as wind regime, sand supply, and surface topography influence the shape of barchans (Howard *et al.*, 1978; Howard and Walmsley, 1985); therefore simplistic geometric models of dune advance (Fisher and Galdies, 1988) are inappropriate. It has also been suggested that variations in surface temperature (Jäkel, 1980) and grain size (Howard *et al.*, 1978) may affect the sand transport rate and maintain a barchan's morphological equilibrium. However, because the formation of a slipface necessitates oversteepening as a result of deposition, small sand mounds do not develop slipfaces. This is because an airstream meeting such a sand patch requires some distance to respond to the increased drag resulting from sand movement before deposition commences (Bagnold, 1941, p. 203). In effect, a barchan's slipface does not extend to the very ends of the horns where they are narrowest, and sand is deflated from these points (Finkel, 1959) (Figure 2.12). Some of this sand is retrieved as the dune advances, but a substantial volume may be lost. So, unless a quantity of sand equal to that lost is supplied to the dune from further upwind, the barchan will ablate. It has been suggested that high rates of sand supply produce less crescentic dunes (Howard *et al.*, 1978) and may therefore result in the development of transverse dune ridges.

There is no theoretical limit to the height a barchan dune may attain under favourable conditions of sand supply. Examples up to 55 m in height have been reported (Simons, 1956). Typically, they average 3.0 to 8.0 m in height (Finkel, 1959; Hastenrath, 1967, 1987; Lettau and Lettau, 1969—in Peru; Long and Sharp, 1964; Inman *et al.*, 1966—in North America; Sarnthein and Walger, 1974—in Mauritania; Fryberger *et al.*, 1984; Watson, 1985, 1986—in Saudi Arabia). Since higher dunes comprise greater volumes of sand, their rate of movement is slower than lower dunes in the same environment. It has also been suggested that a barchan in a low-velocity wind regime will have its crest at the brink of the slipface (Ivanov, 1982; Tsoar, 1985) and will advance at a slower rate than dunes with trapezoidal cross-profiles—with the brink downwind of the crest—which characterize high-velocity wind regimes. Long and Sharp (1964) found a good inverse relationship between barchan dune height and rate of movement (height against log of distance advanced) in Imperial Valley, California, as did Finkel (1959) in Peru. However, the relationship between height and volume is not constant (Norris, 1966) and since volumetric determinations are complex, it is often difficult to predict even relative rates of dune movement. Generally, a dune increasing in volume will decelerate while an ablating dune will accelerate.

Rates of dune advance are quite variable. In some areas, winds blowing from opposite directions create reversing barchans with slipfaces that alternate seasonally—such dunes are often stationary (Lindsay, 1973; Smith, 1981; Machenberg, 1982; Merck, 1983). However, in most areas, unidirectional wind regimes promote net migration of the dunes. Finkel (1959) reported annual rates ranging from 32.3 m for 1.0 m high dunes to 9.2 m/year

Table 2.1 Barchan dune movement rates

Location	Reference	No. of dunes (n)	Mean height (m)	Range in height (m)	Travel distance (m/year)	Range in travel distance (m/year)
Peru	Finkel (1959)	75	3.67	1.0–7.0	15.4	9.2–32.3
	Hastenrath (1967)	42	4.38	1.2–6.0	14.2	9.0–23.0
	Hastenrath (1967)	50	4.12	0.4–6.0	30.7	17.0–56.0
	Hastenrath (1987)	6	4.18	2.4–6.0	22.0	16.0–28.0
	Lettau and Lettau (1969)	114	3.0	—	30.0	—
Mexico	Inman et al. (1966)	—	6.0	—	18.0	—
USA (California)	Long and Sharp (1964)	34	—	—	15.2	7.1–24.4
	Long and Sharp (1964)	47	5.89*	2.7–12.2*	25.0	14.2–39.2
Mauritania	Sarnthein and Walger (1974)	44	8.8	3.0–17.0	30.0	18.0–63.0
Saudi Arabia	Fryberger et al. (1981, 1984)	16	5.7	2.9–12.0	14.8	6.0–28.0
	Watson (1985)	67	8.54†	3.2–25.1†	14.64‡	0.2–44.4‡

*n = 27.
†n = 56.
‡50 weeks.

for 7.0 m high dunes in Peru; the average rate was 15.4 m/year. In the same area, but over a different monitoring period, annual rates from 9.0 to 56 m for dunes from 0.50 to 6.0 m high have been reported (Hastenrath, 1967; Lettau and Lettau, 1969). Hastenrath (1987) noted that over the period between 1964 and 1984 there was a 30 to 40% decrease in sand transport in this region; hence, rates of dune movement are far from constant. The 55 m high Pur-Pur dune of Peru moves about 0.45 to 0.60 m annually (Simons, 1956). In Mexico, rates of 18 m/year for 6.0 m high dunes have been reported (Inman *et al.*, 1966); these are similar to the rates in California—15.4 m/year (Long and Sharp, 1964). However, these same Californian dunes accelerated during a subsequent monitoring period, travelling 25 m annually. In North Africa, rates averaging about 30 m/year for dunes with a mean height of 8.8 m have been determined (Sarnthein and Walger, 1974). In eastern Saudi Arabia, the rate is about 15 m annually (Fryberger *et al.*, 1981, 1984; Watson, 1985). These data are summarized in Table 2.1.

In areas with a great density of barchans, crowding of the dunes, perhaps as a result of an upward gradient which slows their advance, may lead to coalescence (Bagnold, 1951). Barchans coalesce to form aklé dunes and may eventually develop into sinuous, transverse ridges. The potential development of such a form can be seen in Figure 2.12.

The transverse dunes described here may therefore be interrelated—barchans developing from dome dunes and transverse ridges from coalescing barchans. Such transformations are a response to changes in sand supply, wind strength—or direction—or local topography. It should be noted, however, that such morphological evolution is not restricted to transverse forms. In eastern Saudi Arabia, parabolic dunes forming downwind of blowouts induced by devegetation evolve into barchans as sand supply and dune height increase. The opposite trend has also been reported from the White Sands dunefield in New Mexico (McKee, 1983). Holm (1960) described several examples of dune metamorphosis in the deserts of the Arabian Peninsula.

Longitudinal Dunes

These represent some of the largest dune forms. The dune ridges may extend hundreds of kilometres and may reach widths of more than a kilometre and heights of several hundred metres. Typically, heights range from 5.0 to 30 m (Mabbutt, 1977, p. 234), though in some areas they are much greater—50 m in Egypt (Bagnold, 1941, p. 230), 100 m in Libya (Glennie, 1970, p. 95), 150 m in the Rub' al Khālī (Beydoun, 1966) and Namib Desert (Lancaster, 1981a), and 200 m in southern Iran (Gabriel, 1938). The ridges are aligned parallel to each other, the interdune valleys usually being between 1.5 and 3.0 times the width of the dunes (Figure 2.13). However, the relationships

Figure 2.13   Longitudinal dune ridges oriented north–south in the central Namib Desert, Namibia

between dune height and width, between width and spacing, and between height and width are variable (Thomas, 1988). They are probably dependent upon the scale of atmospheric turbulence (Lancaster, 1981b), the rate of sand supply (Mainguet, 1978; Wasson and Hyde, 1983; Thomas, 1988), or sand transmissivity in the interdune areas (Mabbutt, 1984). Often, the processes involved in the development of longitudinal dunes are uncertain (Buckley, 1981). In some cases, the ridges may evolve from the tails of parabolic dunes (Hack, 1941, in the American Southwest; Verstappen, 1968, in the Thar Desert; Twidale, 1981, in the Simpson Desert of Australia) or from coalescing barchans (Bagnold, 1941, pp. 222–4, in the Western Desert of Egypt; Lancaster, 1980, in the Namib Desert; Tsoar, 1984, in the Negev Desert; Kar, 1987, in the Thar Desert). In other cases, the dunes may form through deflation of pans (Lancaster, 1988, in the Kalahari) or interdune depressions (Mainguet, 1978; Buckley, 1981). It does appear, however, that these dune forms grow as sand accumulates at the downwind end of the ridges (Mabbutt and Sullivan, 1968; Lancaster, 1982).

The orientation of the sand ridges is characteristically that of the direction of net sand drift (Bagnold, 1951). Their development is probably associated with dual-directional wind regimes (McKee and Tibbitts, 1964; Breed *et al.*, 1979; Lancaster, 1982; Tsoar, 1982, 1985) with oblique airflow being deflected

Figure 2.14 Snaking crestline of a seif dune in the eastern Rub' al Khālī, United Arab Emirates

in the lee of the dune crests in a net direction parallel to the crestline (Tsoar, 1985; Livingstone, 1988). Though high-velocity winds can move large quantities of sand in short periods (Harmse, 1982), it is likely that relatively low velocity winds blowing over long periods influence longitudinal dune development. This seems to be true in those environments where barchans coalesce to form seif dunes rather than transverse ridges (Lancaster, 1980; Tsoar, 1984). The crests of the dune ridges are characterized by a snaking brinkline about which slipfaces may form on either side if wind directions alternate from both sides of the ridge (Figure 2.14). Generally, winds blowing perpendicular or obliquely to the ridge will modify only the active portion of the crest (Besler, 1977; Livingstone, 1988). While the sand making up these massive dune systems may have moved thousands of kilometres from its source (Fryberger and Ahlbrandt, 1979), the dunes are stable compared with most barchans. However, it has been suggested that lateral migration of longitudinal dunes may be a factor in the misinterpretation of some seif dune deposits in the geological record (Rubin and Hunter, 1985). Such lateral movement produces sedimentary structures resembling those of transverse dunes rather than the cross-bedded patterns found in most linear dunes (McKee and Tibbitts, 1964; McKee, 1982; Tsoar, 1982, 1986).

Within large systems of linear dunes, local wind conditions may lead to the development of star dunes, also known as pyramidal or rhourd dunes. These are stationary forms which grow upwards rather than laterally (Breed and Grow, 1979; Lancaster *et al.*, 1987) reaching heights of over 300 m (Holm, 1960).

THE CONTROL OF MOBILE DUNES

In some ways the problems associated with moving dunes are less severe than those posed by drifting sand. This is because, while sand will drift from any direction that the wind blows at velocities greater than about 5.0 m/s, dune movement requires the mobilization of a large volume of sand and, therefore, the direction of movement is more or less constant on an annual basis. However, whereas installations such as roads, railways, and pipelines can be designed to allow the free movement of sand over them, the encroachment of dunes must be averted.

Relatively few investigations of the methods of combatting the problems posed by encroaching dunes have been undertaken. Those by Kerr and Nigra (1952) in Saudi Arabia and Finkel (1959) in Peru remain the cornerstones in the field. Several of Finkel's main conclusions merit reiteration here: first, it is essential to acknowledge that the control of wind-blown sand may have no effect whatsoever on the rate of dune movement; second, dunes cannot be tolerated in the same way as blowing sand might; third, source areas are often very distant, so stabilization of the sand source may prove impractical or impossible; fourth, ditches are useless since they fill with sand almost immediately while having little effect on dune morphodynamics; fifth, mechanical disturbance is often ineffective; sixth, physical barriers are also ineffective; seventh, vegetative stabilization techniques are usually impractical because they require irrigation; and eighth, techniques that reduce the height or volume of the dunes may merely aggravate the problem by increasing the rate of advance.

Notwithstanding these observations, the three main procedures for combatting the problems posed by dune encroachment are:

(a) removal of the dunes;
(b) dissipation of the dunes; and
(c) immobilization of the dunes.

The selection of one of these options depends on the type of installation being protected, the distance of the dunes from the installation, and, to some extent, the size of the dune.

Removal of the Dunes

The only practical method of dune removal is mechanical excavation and transportation to a new location. The costs are high since a 6.0 m high dune may incorporate 20 000 to 25 000 m^3 of sand, weighing between 30 000 and 45 000 tonnes depending on its volumetric porosity. Often the removal of dunes is only practical when the sand can be used as fill or ballast, though the difficulty in compacting aeolian sand—especially in the absence of water—frequently precludes its use. In Central Asia, however, dune sand is used in the manufacture of concrete (Choshchshiev and Vekilov, 1985) and cements for road surfacing (Milyavskaya and Kalandarishvili, 1985).

The partial removal of mobile dunes should not be undertaken since it could lead to increased mobility of the residual dune. Great care must be exercised in the implementation of all of the following dune control measures since the creation of smaller, more mobile dunes may exacerbate the problems. One of the few instances when a reduction in dune height may suffice is when high dunes encroach upon aerial powerlines—the smaller dunes can pass beneath the cables.

Dissipation of the Dunes

Mobile sand dunes can be destroyed and the sand removed by natural processes. This is practical only in areas where drifting sand will not create a hazard. The dissipation of a mobile dune is achieved by disrupting its aerodynamic profile. This can be accomplished in three ways.

Reshaping In very simple terms, a barchan moves through a process of sand transport from the windward side up the 13 to 22% slope, followed by deposition in the area of reduced wind strength beyond the crest, that is on the slipface. In effect, dune movement involves deflation of the windward side of the dune and accretion of the slipface. Hence, if the slipface is removed mechanically and prevented from reforming, sand will flow beyond the dune (Figures 2.11 and 2.12).

The morphology of the windward portion of the dune also has a significant effect on dune mobility since it influences sand-transport rates. As Lai and

Wu (1978, p. 24) pointed out, 'a dune with a slow increase in local slope is more stable than a dune with a sharp increase in slope near the peak', even if the dunes are of the same height and length (Figure 2.2). Hence, reshaping of both the windward and leeward portions may be essential. However, any modification will be temporary since the sand mound will tend to evolve towards the optimum aerodynamic form. The surface of the reshaped dune must be stabilized in some way if the treatment is to be permanent.

Trenching   A more permanent method of dissipating a mobile dune involves disrupting the aerodynamic form by excavating a trench parallel to the direction of movement, through the axis of the dune. The trench will alow un restricted airflow through the centre of the dune and remove sand from the core. Panelling has also been employed to direct the airflow on portions of the dune, thereby creating small blowouts which disrupt the profile (Clements *et al.*, 1963). Since it is possible that the deflating sand could form a mobile parabolic dune, or the two portions of the bisected barchan may be shaped into smaller dunes and continue to advance, this technique should be adopted only well upwind of sensitive installations.

Surface treatment   This involves surface stabilization of the arms of a barchan allowing deflation of the central portion (Figure 2.11). The stabilization can be undertaken using oil, chemicals, gravel coatings, or fences. The technique has the benefit of not only stabilizing the mobile dune but also reducing sand supply from the horns of the dune to other dunes further downwind (Finkel, 1959). However, once again the technique's value is limited owing to the problems caused by deflating sand. ARAMCO has applied the technique successfully on barchans encroaching upon powerlines (Trossel, 1981).

In places where the liberation of large quantities of sand from an ablating dune will not pose a hazard, the introduction of pebbles or other large particles to the windward surface of the dune—as few as one or two per square metre may suffice—can disrupt the sand flow over the surface, eventually destroying the dune (Barclay, 1917; Holm, 1960; Lettau and Lettau, 1969).

Immobilization of the Dunes

In areas where mobile dunes pose an immediate threat, destructive techniques are inappropriate. Finkel (1959) stressed three basic principles upon which dune control under such conditions should be undertaken: first, if possible the sand source should be identified and stabilized; second, the smaller, more mobile dunes require priority treatment; and, third, since the horns of barchans supply sand to dunes downwind, they should be preferentially stabilized. If source area stabilization or complete removal of the dunes is not viable on economic grounds, several methods of dune immobilization can be employed to avert the threat of encroachment.

Trimming Dunes can be reshaped, as described above, in order to alter their aerodynamic form and retard their movement. However, if the surface of the reshaped dune is not treated to prevent deflation, the area downwind will experience pronounced sand drift. While dune trimming is a prerequisite for dune immobilization by other methods, it does not in itself constitute an effective method of long-term stabilization.

Surface treatment The type of treatment employed is not significant; crude oil, cutback asphalt, and chemicals have all been used in Saudi Arabi. The armouring of a dune surface with aggregate or artificial coarse-grained particles can prevent sand entrainment (Holm, 1960). The method is useful in areas where aggregate is readily available and where environmental considerations preclude the use of oil or chemical stabilizers.

The size and morphology of dunes affect the grain-size characteristics of the sand upon them by directly influencing variations in shear stress over the surface (Tsoar, 1986; Watson, 1986; McArthur, 1987; Hartmann and Christiansen, 1988). This must be taken into consideration when dunes are reshaped or are stabilized using surface treatment techniques. Conversely, the variations in grain size within sand seas and from erg to erg may also influence dune morphology (Warren, 1972; Wilson, 1973; Lancaster, 1983). Over the Arabian Peninsula, for example, there are marked differences in the average size of dune sands (Table 2.2 and Figure 2.15). On a local scale, variations in median grain size result from changes in dune dimensions; the sand at the crests of high dunes is often finer and better sorted than sand from smaller dunes. At the regional scale, wind regime and wind strength, as well as the nature of the source materials, are preeminent factors determining the different grain-size characteristics of dune sands (Wasson and Hyde, 1983; Thomas, 1987). It is just as important to take these factors into account when implementing sand control measures as it is to tailor the schemes to the broader environmental parameters (Li and Khudaiyarova, 1987).

ARAMCO and the Saudi Arabian Ministry of Communications have frequently resorted to mantling completely the windward side of barchan

Table 2.2 Arabian Peninsula dune sand grain-size data

Sample*	Location	Dune form	Height (m)	Md† (μm)
1	Jafurah (north)	Barchan	30	300
2	Jafurah (north)	Barchan	20	312
3	Jafurah (north)	Barchan	12	328
4	Jafurah (north)	Barchan	2	308
5	Jafurah (south)	Barchan	6	336
6	Jafurah (south)	Barchan	8.5	296
7	Jafurah (south)	Barchan	8	293
8	Jafurah (south)	Barchan	15	312
9	Jafurah (south)	Barchan	20	324
10	Dahna (central)	Seif	25	227
11	Dahna (central)	Seif	5	285
12	Dahna (south)	Transverse ridge	6	332
13	Dahna (south)	Transverse ridge	5	304
14	Nefud ath Thuwayrat	Seif	55	293
15	Nefud as Sirr	Seif	5	300
16	Nefud Qunayfidhah	Seif	N/A	289
17	Rub al Khali (east)	Seif	40	293
18	Rub al Khali (central)	Seif	N/A	271
19	Rub al Khali (west)	Barchan	2	285
20	Rub al Khali (west)	Seif	51	230
21	Rub al Khali (west)	Star dune	28	236

All sand samples were collected from the brinks of barchans and transverse ridges, and from the crests of seifs and the star dune.
*Sample numbers refer to those shown on Figure 2.15.
†Median grain size.

dunes with crude oil or cutback asphalt (Figure 2.4). This immobilizes the sand body but enhances sand grain saltation up the windward slope, resulting in rapid accumulation of sand in the lee of the dune. Trossel (1981) held that the accumulation zone will not extend more than a distance equal to 20 times the height of the dune from the original dune. However, the sand accumulation is unstable and a secondary, mobile dune may evolve. An alternative procedure involves shaping the dune prior to surface treatment in order to reduce the sand-trapping effect of the slipface. Provided that the leeward slope does not exceed 1:10 (13%), sand will not be deposited there (Figure 2.11). This procedure can be employed only upwind of installations that are designed to allow the unhindered through-flow of wind-blown sand.

An important modification to the technique of surface treatment is the creation of stabilized strips rather than complete mantling of the surface. This is increasingly widespread throughout Saudi Arabia since it requires significantly less material. Strip stabilization involves laying 2.0 m wide bands of

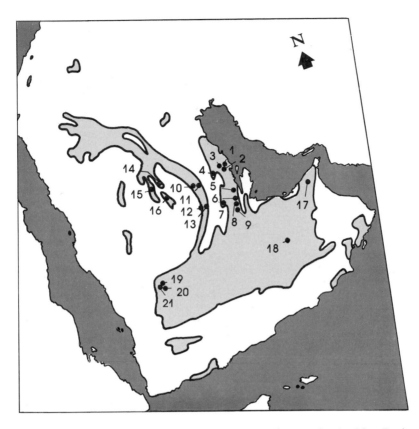

Figure 2.15   Distribution of sand seas (lightly shaded) over the Arabian Peninsula. Numbers refer to dune sampling sites for which granulometric data are presented in Table 2.2

surface stabilizer perpendicular to the direction of dune movement (Figure 2.16). The strips are separated by untreated areas twice their width. Initially, sand is deflated from the dunes by scouring of the untreated bands but, if the strips are not undercut, scouring will eventually cease and the dunes will be stabilized. If the dunes are shaped to reduce sand accumulation on their leeward sides, secondary dune formation is inhibited. Since scouring of the untreated portions is inevitable, it is crucial to select an oil, asphalt, or chemical stabilizer which penetrates the sand surface and which is not easily destroyed by undercutting (Figure 2.17) or sand-blasting (Azizov and Atabaev, 1987). The technique has been employed using pressure injection of an emulsified cutback asphalt (Figure 2.18) along the road from Al-Muzahimiyah/Al-Quwayiyah to Zalim, 100 km west of Riyadh (BMMK and Partners, no date) (Figures 2.1, 2.16, 2.17 and 2.18). While dune encroach-

Figure 2.16   Strip stabilization in the Nefud Qunayfidhah along the road from
Al-Muzahimiyah to Zalim

Figure 2.17   Destruction of stabilized strips by undercutting as a result of aeolian
scour, and by dune encroachment

Figure 2.18   Close-up of one of the stabilized strips shown in Figures 2.16 and 2.17. Surface penetration of 28 mm for the emulsified cutback asphalt–latex mixture was achieved through pressure injection

ment in this area is not as severe as in the Jafurah sand sea, similar techniques implemented between Al-Hasa and Salwah (Figure 2.1) have had marked success in immobilizing dunes for periods of five years and more.

Fences   In areas where protection is required from drifting sand as well as from dune encroachment, the most effective stabilization policy involves the erection of sand fences to supplement dune immobilization. The fences can be employed in two ways:

(a) If the dunes are more than about one kilometre from an installation, fences can be emplaced upwind of the dune to trap incoming sand. During the normal process of barchan movement some sand is deflated from the horns and is lost to the dune system. In addition, some sand will fill depressions in the surface over which the dune is migrating (Fryberger *et al.*, 1988, p. 33). Hence, if the sand supply to the dune is cut off, the dune will gradually decrease in size. In China, barchans have been dissipated by planting grasses in the interdune areas to trap the dune sand (Liu Shu, 1986). The distance the dune will travel before it disappears completely depends on the size of the dune, its rate of movement, and the rate of sand loss. Sand loss from 20 m high barchans travelling across a sabkha near Dammam (Figures 2.1 and 2.12) have been estimated from aerial surveys of the extent of plumes of deflating sand and measurements of the volume of this sand per unit area on the sabkha surface. The volume lost from each dune is about 2000 m^3/year, about 2.0 to 3.0% of the dune's total volume. Assuming that this figure and the rate of movement remain constant—though Norris (1966) found that ablating dunes accelerate—a 10 m high dune travelling at 15 m/year would be less than 1.0 m high in about 60 years having moved 900 m.

The greatest problem with this approach to dune control is that the fences that inhibit sand supply will themselves accumulate sand. Dunes built by sand fencing are permanent features that must be periodically refenced as sand accumulates. The siting of fences is critical because over a long period very high dunes can accrete. Kerr and Nigra (1952) estimated that in the Dhahran area of Saudi Arabia, dunes would be about 7.0 m high after 10 years, about 11 m after 20 years, 16 m after 50 years, and 30 m after 200 years.

(b) If the encroaching dunes present an immediate threat, they must be immobilized using surface treatment techniques. If protection from drift-

Figure 2.19   Destruction of palm-leaf fences by aeolian scour on the flank of a low dune

Figure 2.20 Aeolian scour at the base of a fabric sand-fence erected on a dune; at the time of installation the lowest horizontal strip was level with the sand surface

ing sand is also necessary, fences should be constructed upwind of the dune. Their precise location can be determined by field monitoring of the lines of sand movement or identification of these paths from aerial photographs. The rate of growth of artificial dunes and the scheduling of fence replacement can be calculated using field measurements of the rate of sand transport. Under severe conditions, and where sand drift must be minimized, fences can be erected on the dune itself. Manohar and Bruun (1970) used wind tunnel simulations to determine the optimal location of fences relative to the dune crest, taking fence height and wind velocity into consideration. If the fences are erected on an unconsolidated surface—especially on a dune—scouring may occur (Figures 2.19 and 2.20) as a result of jetting of the airflow. In such cases, the surface adjacent to the fence must be stabilized using some type of surface treatment.

SUMMARY AND CONCLUSIONS

In desert regions with strong winds and an abundant supply of unconsolidated, fine-grained sediments, engineers are often confronted with two main problems of sand control—drifting sand and mobile dunes. While it is possible to prevent sand drift using fences and to halt dunes by treating their surfaces, implementation of these procedures initiates other problems. In these regions, it must be acknowledged that there is an almost limitless supply of drifting sand. This cannot be halted except on a very local scale; therefore it must be accommodated. Based upon empirical data, sand drift rates in the Eastern Province of Saudi Arabia are up to about 30 m^3/m width/year and the average dune movement is about 15 m/year. While these rates are very high, the region has the advantage of experiencing unidirectional sand drift and dune movement. In effect, engineers need to tackle problems of sand originating from only one direction. In desert areas where sand drift directions are not so uniform, the difficulties are multiplied.

The problems caused by wind-blown sand can be tackled in four ways: (a) by promoting its deposition, using ditches, fences, or tree belts located upwind of the area needing protection; (b) by enhancing its transport through the area, using streamlining techniques, creating a smooth texture over the land surface, or by erecting panels to direct the airflow; (c) by reducing the supply of sand upwind, using surface stabilization techniques, fences, or vegetation; (d) by deflecting the sand using fences or tree belts.

The most effective long-term solution is to design the facilities to allow the free movement of sand across them. However, since this is not always feasible—buildings, for example, will almost always create areas of reduced wind strength wherein sand will accumulate—fences can be erected upwind to trap the moving sand. The employment of fences or vegetation belts constitutes a commitment to dune building. The accumulation of sand will continue indefinitely and the fences or trees must be maintained accordingly.

Moving dunes can be dealt with in three ways: (a) by removing them mechanically; (b) by dissipating them using reshaping, trenching, or surface stabilization techniques; (c) by immobilizing them through altering their aerodynamic form, by surface stabilization or by using fences.

The best remedy for the problem—the removal of the dunes—is not always practical or viable. In effect, they must be either dissipated or immobilized. Because of the threat from drifting sand, destructive procedures can be implemented only with dunes at some distance from a facility. Those dunes that pose an immediate threat must be immobilized by treating their surfaces. The way in which this is undertaken depends upon economic and environmental considerations. At present, in Saudi Arabia, the most viable approach involves strip stabilization of the dune surface using oil-based substances.

The success of any sand control scheme is dependent upon detailed monitoring of sand flow and dune dynamics (Watson, 1985; Jones *et al.*, 1986; Cherednichenko, 1987a). This can be accomplished by mapping geomorphological features or characteristic vegetation types (Lyasovskaya, 1985) from aerial photographs or satellite imagery. The increasing availability of sequential satellite imagery can now provide geomorphologists and engineers with data on the rate of desertification in many remote arid regions (Vinogradov and Kulik, 1987). Such information may be supplemented with meteorological data in order to estimate potential rates of sand flow. However, the acquisition of quantitative data on rates of sand drift requires field-based studies using sand traps.

A major consideration in the implementation of any sand control scheme is the necessity to minimize disruption of naturally stable portions of the landscape. The destruction of vegetation or breaching of desert pavement and salt crusts can lead to rapid scouring of unconsolidated materials (Arnagel' dyev and Kostyukovsky, 1982; Marston, 1986); indeed, devegetation of sand sheets may be the main factor in the formation of some mobile dunes (Hack, 1941; Tsoar and Møller, 1986; Castel, 1988). A disturbed desert landscape may require hundreds or thousands of years to recover (Webb *et al.*, 1988). In most cases, large-scale sand control programmes which involve the integration of several different control techniques require a stratified approach employing various intensities of sand control in different zones depending on the severity of the hazard posed by sand movement (Cherednichenko, 1987b).

REFERENCES

Adriani, M. J. and Terwindt, J. H. J. (1974). Sand stabilization and dune building, *Rijkswaterstaat Communications*, **19**, 68 pp.
Ahmad, R. (1986). Stabilization and afforestation of sand dunes through biosaline culture technique, in *Physics of Desertification* (Eds. F. El-Baz and M. H. A. Hassan), pp. 109–17, Martinus Nijhoff, Dordecht.
Anderson, R. S. (1987). A theoretical model for aeolian impact ripples, *Sedimentology*, **34**, 943–56.
Anton, D. (1983). Modern eolian deposits of the Eastern Province of Saudi Arabia, in *Eolian Sediments and Processes* (Eds. M. E. Brookfield and T. S. Ahlbrandt), pp. 365–78, Elsevier, Amsterdam.
Aripov, E. A. and Nuryev, B. N. (1982). Classification of chemical ameliorants used for stabilizing shifting sands, *Problems of Desert Development*, No. 6, 58–60.
Arnagel'dyev, A. and Kostyukovsky, V. I. (1982). Protection of catchment plots and installations from sand drift at takyrs, *Geomorfologiya*, No. 1, 39–43.
Arnagel'dyev, A. and Kurbanov, O. R. (1984). Protection of industrial facilities from sand drifts and deflation in southwest Turkmenistan, *Problems of Desert Development*, No. 2, 84–8.
Azizov, A. and Atabaev, B. (1987). Influence of windborne sand on erosion resistance of soils treated with various preparations, *Problems of Desert Development*, No. 2, 85–90.
Bagnold, R. A. (1941). *The Physics of Blown Sand and Desert Dunes*, Methuen, London.
Bagnold, R. A. (1951). Sand formations of southern Arabia, *Geographical Journal*, **117**, 78–86.

Barclay, W. S. (1917). Sand-dunes in the Peruvian desert, *Geographical Journal*, **49**, 53–6.

Besler, H. (1977). Untersuchungen in der Dünen-Namib (Südwestafrika). Vorläufige Ergebnisse des Forschungsaufenthaltes, 1976, *Journal of the South West Africa Scientific Society*, **31**, 33–64.

Besler, H. (1984). Verschiedene Typen von Reg, Dünen und kleinen Ergs in der algerischen Sahara, *Die Erde*, **115**, 47–79.

Beydoun, Z. R. (1966). Geology of the Arabian Peninsula. Eastern Aden Protectorate and part of Dhufar, United States Geological Survey, Professional Paper 560H, 1–49.

Bisal, F. and Hsieh, J. (1966). Influence of moisture on erodibility of soil by wind, *Soil Science*, **102**, 143–6.

BMMK and Partners (no date). Al-Muzahimiyah–Al-Quwayiyah–Zalim Road, Section 4. The control of wind blown sand adjacent to highways, unpublished report to Saudi Arabian Ministry of Communications, Technical Administration for Roads and Ports, 16 pp.

Breed, C. S. and Grow, T. (1979). Morphology and distribution of dunes and sand seas observed by remote sensing, in *A Study of Global Sand Seas* (Ed. E. D. McKee), pp. 253–302, United States Geological Survey, Professional Paper 1052.

Breed, C. S., Fryberger, S. G., Andrews, S., McCauley, C., Lennartz, F., Gebel, D. and Horstman, K. (1979). Regional studies of sand seas using LANDSAT (ERTS) imagery, in *A Study of Global Sand Seas* (Ed. E. D. McKee), pp. 305–97, United States Geological Survey, Professional Paper 1052.

Brown, R. L. (1948). Permanent coastal dune stabilization with grasses and legumes, *Journal of Soil and Water Conservation*, **3**, 69–74.

Brown, R. L. and Hafenrichter, A. L. (1962). Stabilizing sand dunes on the Pacific Coast with woody plants, United States Department of Agriculture, Soil Conservation Service—Miscellaneous Publication **892**, 18 pp.

Buckley, R. (1981). Central Australian sand ridges, *Journal of Arid Environments*, **4**, 91–101.

Castel, I. I. Y. (1988). A simulation model of wind erosion and sedimentation as a basis for management of a drift sand area in The Netherlands, *Earth Surface Processes and Landforms*, **13**, 501–9.

Chepil, W. S. and Woodruff, N. P. (1963). The physics of wind erosion and its control, *Advances in Agronomy*, **15**, 211–302.

Cherednichenko, V. P. (1987a). On geomorphological studies of sand deposits with a view to engineering, *Geomofologiya*, no. 4, 21–6.

Cherednichenko, V. P. (1987b). Geomorphological zoning of deserts of northern Turkmenistan for engineering purposes, *Problems of Desert Development*, no. 4, 26–32.

Choshchshiev, K. C. and Vekilov, D. (1985). Lightweight concrete using barchan sands, *Problems of Desert Development*, no. 2, 93–4.

Clements, T., Stone, R. O., Mann, J. F. Jr and Eymann, J. L. (1963). A study of windborne sand and dust in desert areas, United States Army, Natick Laboratory (Natick, Mass.), Technical Report **ES-8**, 61 pp.

Davis, J. H. (1975). Stabilization of beaches and dunes by vegetation in Florida, State University System of Florida, Sea Grant Program 7, 52 pp.

Evans, J. R. (1962). Falling and climbing sand dunes in the Cronese ('Cat') Mountain area, San Bernardino County, California, *Journal of Geology*, **70**, 107–13.

Fink, L. K. and Smith, M. M. (1974). Stabilizing the Ogunquit dunes, is it best for the beach?, *Maine Environment, Bulletin of the Natural Resources Council of Maine*, **1974**, 1–5.

Finkel, H. J. (1959). The barchans of southern Peru, *Journal of Geology*, **67**, 614–47.

Fisher, P. F. and Galdies, P. (1988). A computer model for barchan dune movement, *Computers and Geosciences*, **14**, 229–53.

Fryberger, S. G. and Ahlbrandt, T. S. (1979). Mechanisms for the formation of eolian sand seas, *Zeitschrift für Geomorphologie, N.F.*, **23**, 440–60.

Fryberger, S. G., Al-Sari, A. M., Clisham, T. J., Risvi, S. R. and Al-Hinai, K. G. (1981). Dune advance, sand drift rates and wind regime, Eastern Province, Saudi Arabia, and applications for control of wind-driven sand, unpublished report, Research Institute, Univesity of Petroleum and Minerals, Dhahran, Saudi Arabia.

Fryberger, S. G., Al-Sari, A. M. and Clisham, T. J. (1983). Eolian dune, interdune, sand sheet, and siliciclastic sabkha sediments of an offshore prograding sand sea, Dhahran area, Saudi Arabia, *Bulletin of the American Association of Petroleum Geologists*, **67**, 280–312.

Fryberger, S. G., Al-Sari, A. M., Clisham, T. J., Risvi, S. A. R. and Al-Hinai, K. G. (1984). Wind sedimentation in the Jafurah sand sea, Saudi Arabia, *Sedimentology*, **31**, 413–31.

Fryberger, S. G., Schenk, C. J. and Krystinik, L. F. (1988). Stokes surfaces and the effects of near-surface groundwater-table on aeolian deposition, *Sedimentology*, **35**, 21–41.

Gabriel, A. (1938). The southern Lut and Iranian Baluchistan, *Geographical Journal*, **92**, 195–210.

Gabriels, D., de Boodt, M. and Minjauw, D. (1974). Dune sand stabilization with synthetic soil conditioners; a laboratory experiment, *Soil Science*, **118**, 332–8.

Garés, P. A., Nordstrom, K. F. and Psuty, N. P. (1979). Coastal dunes: their function, delineation and management, New Jersey Department of Environmental Protection, Division of Coastal Resources/Rutgers University Center for Coastal and Environmental Studies, New Brunswick, N.J.

Garés, P. A., Nordstrom, K. F. and Psuty, N. P. (1980). Delineation and implementation of a dune management district, *Proceedings of the Coastal Zone '80 Conference, Hollywood, Fla.*, pp. 1269–88.

Gillette, D. A. and Walker, T. R. (1977). Characteristics of airborne particles produced by wind erosion of sandy soil, High Plains of West Texas, *Soil Science*, **123**, 97–110.

Gillette, D. A., Adams, J., Endo, A., Smith, D. and Kihl, R. (1980). Threshold velocities for input of soil particles into the air by desert soils, *Journal of Geophysical Research*, **85**, 5621–30.

Gillette, D. A., Adams, J., Muhs, D. and Kihl, R. (1982). Threshold friction velocities and rupture moduli for crusted desert soils for the input of soil particles into the air, *Journal of Geophysical Research*, **87**, 9003–15.

Glennie, K. W. (1970). *Desert Sedimentary Environments*, Elsevier, Amsterdam.

Gupta, J. P. (1979). Some observations on the periodic variations of moisture in stabilized and unstabilized dunes of the Indian Desert, *Journal of Hydrology*, **41**, 153–6.

Hack, J. T. (1941). Dunes of western Navajo Country, *Geographical Review*, **31**, 240–6.

Harmse, J. T. (1982). Geomorphologically effective winds in the northern part of the Namib sand desert, *South African Geographer*, **10**, 43–52.

Hartmann, D. and Christiansen, C. (1988). Settling-velocity distributions and sorting processes on a longitudinal dune: a case study, *Earth Surface Processes and Landforms*, **13**, 649–56.

Hastenrath, S. L. (1967). The barchans of the Arequipa region, southern Peru, *Zeitschrift für Geomorphologie, N.F.*, **11**, 300–31.

Hastenrath, S. (1987). The barchan dunes of southern Peru revisited, *Zeitschrift für Geomorphologie, N.F.*, **31**, 167–78.

Hawk, V. B. and Sharp, W. C. (1967). Sand dune stabilization along the North Atlantic coast, *Journal of Soil and Water Conservation*, **22**, 143–6.

Hesp, P. A. (1981). The formation of shadow dunes, *Journal of Sedimentary Petrology*, **51**, 101–12.

Hidore, J. J. and Albokhair, Y. (1982). Sand encroachment at Al-Hasa Oasis, Saudi Arabia, *Geographical Review*, **72**, 350–6.

Holm, D. A. (1960). Desert geomorphology in the Arabian Peninsula, *Science*, **132**, 1369–79.

Howard, A. D. (1985). Interaction of sand transport with topography and local winds in the northern Peruvian coastal desert, *Memoirs of the Department of Theoretical Statistics, Institute of Mathematics, University of Aarhus*, **8**, 511–43.

Howard, A. D. and Walmsley, J. L. (1985). Simulation model of isolated dune sculptured by wind, *Memoirs of the Department of Theoretical Statistics, Institute of Mathematics, University of Aarhus*, **8**, 377–91.

Howard, A. D., Morton, J. B., Gad-el-Hak, M. and Pierce, D. B. (1978). Sand transport model of barchan dune equilibrium, *Sedimentology*, **25**, 307–38.

Hunter, R. E. (1985). A kinematic model for the structure of lee-side deposits, *Sedimentology*, **32**, 409–22.

Inman, D. C., Ewing, G. C. and Corliss, J. B. (1966). Coastal sand dunes of Guerrero Negro, Baja California, Mexico, *Bulletin of the Geological Society of America*, **77**, 787–802.

Ivanov, A. P. (1982). Travel speed of a barchan in relation to its profile, height and wind speed, *Problems of Desert Development*, No. 6, 61–3.

Jagschitz, J. A. and Wakefield, R. C. (1971). How to build and save beaches and dunes, Agricultural Experimental Station Bulletin 408, University of Rhode Island Marine Leaflet Series 4, 12 pp.

Jäkel, D. (1980). Die Bildung von Barchanen in Faya-Largeau/Rep. du Tchad, *Zeitschrift für Geomorphologie, N.F.*, **24**, 141–59.

Johnson, D. H., Kamal, M. R., Pierson, G. O. and Ramsay, J. B. (1978). Sabkhas in eastern Saudi Arabia, in *Quaternary Period in Saudi Arabia* (Eds. S. S. Al-Sayari and J. G. Zötl), pp. 84–93, Springer-Verlag, Wien.

Jones, D. K. C., Cooke, R. U. and Warren, A. (1986). Geomorphological investigation, for engineering purposes, of blowing sand and dust hazard, *Quarterly Journal of Engineering Geology, London*, **19**, 251–70.

Jungerius, P. D., Verheggen, A. J. T. and Wiggers, A. J. (1981). The development of blowouts in 'de Blink', a coastal dune area near Noordwijkerhout, The Netherlands, *Earth Surface Processes and Landforms*, **6**, 375–96.

Kar, A. (1987). Origin and transformation of longitudinal sand dunes in the Indian Desert, *Zeitschrift für Geomorphologie, N.F.*, **31**, 167–78.

Kerr, R. O. and Nigra, J. O. (1952). Eolian sand control, *Bulletin of the American Association of Petroleum Geologists*, **36**, 1541–73.

Khodzhaev, C. (1983). Dynamics of barchan sands in the Karakum Canal Zone, *Problems of Desert Development*, No. 4, 53–8.

Knutson, P. L. (1977). Planting guidelines for dune creation and stabilization, United States Army Corps of Engineers, Coastal Engineering Research Centre, Technical Aid, 77(4), 26 pp.

Knutson, P. L. (1980). Experimental dune restoration and stabilization, Nauset Beach, Cape Cod, Massachusetts, United States Army Corps of Engineers, Coastal Engineering Research Center, Technical Paper 80-5, 42 pp.

Lai, R. J. and Wu, J. (1978). Wind erosion and deposition along a coastal sand dune, University of Delaware, Sea Grant College Program DEL-SG-10-78, 26 pp.

Lancaster, N. (1980). The formation of seif dunes from barchans—supporting evidence for Bagnold's model from the Namib Desert, *Zeitschrift für Geomorphologie, N.F.*, **24**, 160–7.

Lancaster, N. (1981a). Grain size characteristics of Namib Desert linear dunes, *Sedimentology*, **28**, 115–22.

Lancaster, N. (1981b). Aspects of the morphometry of linear dunes of the Namib Desert, *South African Journal of Science*, **77**, 366–8.

Lancaster, N. (1982). Linear dunes, *Progress in Physical Geography*, **6**, 475–504.

Lancaster, N. (1983). Controls of dune morphology in the Namib sand sea, in *Eolian Sediments and Processes* (Eds. M. E. Brookfield and T. S. Ahlbrandt), pp. 261–89, Elsevier, Amsterdam.

Lancaster, N. (1985). Variations in wind velocity and sand transport on the windward side of desert sand dunes, *Sedimentology*, **32**, 581–93.

Lancaster, N. (1988). Development of linear dunes in the southwestern Kalahari, Southern Africa, *Journal of Arid Environments*, **14**, 233–44.

Lancaster, N., Greeley, R. and Christensen, P. R. (1987). Dunes of the Gran Desierto sand sea, Sonora, Mexico, *Earth Surface Processes and Landforms*, **12**, 277–288.

Lettau, K. and Lettau, H. (1969). Bulk transport of sand by the barchans of Pampa de la Joya in southern Peru, *Zeitschrift für Geomorphologie, N.F.*, **13**, 182–95.

Li, R. A. and Khudaiyarova, B. (1987). Particle size distribution of eolian sands of the eastern part of the low-lying Karakum, *Problems of Desert Development*, No. 4, 79–81.

Lindsay, J. F. (1973). Reversing barchan dunes in Lower Victoria Valley, Antarctica, *Bulletin of the Geological Society of America*, **84**, 1799–806.

Liu Shu (1986). Basic ways of securing mobile sands in China, *Problems of Desert Development*, No. 3, 78–81.

Livingstone, I. (1988). New models for the formation of linear sand dunes, *Geography*, **73**, 105–15.

Logie, M. (1981). Wind tunnel experiments on dune sands, *Earth Surface Processes and Landforms*, **6**, 364–74.

Long, J. T. and Sharp, R. P. (1964). Barchan dune movement in Imperial Valley, California, *Bulletin of the Geological Society of America*, **75**, 149–56.

Lyasovskaya, L. M. (1985). The application of vegetation indicators to geological-engineering surveying in the Karakalpak Ustyurt, *Problems of Desert Development*, No. 1, 91–4.

Lyles, L. and Schrandt, R. L. (1972). Wind erodibility as influenced by rainfall and soil salinity, *Soil Science*, **114**, 367–72.

Mabbutt, J. A. (1977). *Desert Landforms*, MIT Press, Cambridge, Mass.

Mabbutt, J. A. (1984). Discussion of 'Factors determining desert dune type', *Nature*, **309**, 91–2.

Mabbutt, J. A. and Sullivan, M. E. (1968). The formation of longitudinal dunes: evidence from the Simpson Desert, *Australian Geographer*, **10**, 483–7.

McArthur, D. S. (1987). Distinctions between grain-size distributions of accretion and encroachment deposits in an inland dune, *Sedimentary Geology*, **54**, 147–63.

McFadden, L. D., Wells, S. G. and Jercinovich, M. J. (1987). Influences of eolian and pedogenic processes on the origin and evolution of desert pavements, *Geology*, **15**, 504–8.

Machenberg, M. D. (1982). Sand dune migration in Monahans Sandhills State Park, Texas, *Geological Society of America, Abstracts with Programs*, **14**, 116.

McKee, E. D. (1979a). Introduction to a study of global sand seas, in *A Study of Global Sand Seas* (Ed. E. D. McKee), pp. 1–19, United States Geological Survey, Professional Paper 1052.

McKee, E. D. (1979b). Sedimentary structures in dunes, in *A Study of Global Sand Seas* (Ed. E. D. McKee), pp. 83–134, United States Geological Survey, Professional Paper 1052.

McKee, E. D. (1982). Sedimentary structures in dunes of the Namib Desert, South West Africa, Geological Society of America, Special Paper 188, 64 pp.

McKee, E. D. (1983). Eolian sand bodies of the world, in *Eolian Sediments and Processes* (Eds. M. E. Brookfield and T. S. Ahlbrandt), pp. 1–25, Elsevier, Amsterdam.

McKee, E. D. and Tibbitts, G. C. (1964). Primary structures of a seif dune and associated deposits in Libya, *Journal of Sedimentary Petrology*, **34**, 5–17.

Mader, D. and Yardley, M. J. (1985). Migration, modification and merging in aeolian systems and the significance of the depositional mechanisms in Permian and Triassic dune sands of Europe and North America, *Sedimentary Geology*, **43**, 85–218.

Mainguet, M. (1978). The influence of Trade Winds, local air-masses and topographic obstacles on the aeolian movement of sand particles and the origin and distribution of dunes and ergs in the Sahara and Australia, *Geoforum*, **9**, 17–28.

Manohar, M. and Bruun, P. (1970). Mechanics of dune growth by sand fences, *Dock and Harbour Authority*, **51**, 243–52.

Marston, R. A. (1986). Maneuver-caused wind erosion impacts, south central New Mexico, in *Aeolian Geomorphology* (Ed. W. G. Nickling), pp. 273–90, Allen and Unwin, Boston, Mass.

Mercer, A. G. and Haque, M. I. (1973). Ripple profiles modelled mathematically, *Proceedings of the American Society of Civil Engineers, Journal of the Hydraulics Division*, **99**, 441–59.

Merck, G. (1983). Dune form and structure at Great Sand Dunes National Monument, CO, *Geological Society of America, Abstracts with Programs*, **15**, 227.

Milyavskaya, M. B. and Kalandarishvili, A. G. (1985). Reinforcing sandy soil by means of surface-active agents, *Problems of Desert Development*, No. 6, 95–7.

Mirakhmedov, M. (1983). Stabilization of mobile sands with heavy crude, *Problems of Desert Development*, No. 1, 88–9.

Mulligan, K. R. (1988). Velocity profiles measured on the windward slope of a transverse dune, *Earth Surface Processes and Landforms*, **13**, 573–582.

Nickling, W. G. (1984). The stabilizing role of bonding agents on the entrainment of sediment by wind, *Sedimentology*, **31**, 505–10.

Nickling, W. G. and Ecclestone, M. (1981). The effects of soluble salts on the threshold shear velocity of fine sand, *Sedimentology*, **28**, 505–10.

Nordstrom, K. F. and Psuty, N. P. (1980). Dune district management: a framework for shorefront protection and land use control, *Coastal Zone Management Journal*, **7**, 1–23.

Norris, R. M. (1966). Barchan dunes of Imperial Valley, California, *Journal of Geology*, **74**, 292–306.

Nuryev, B. N., Babaev, M. G., Aripov, E. A. and Kapurov, K. (1985). Stabilization of mobile sands with an asphalt emulsion paste, *Problems of Desert Development*, No. 2, 90–2.

Ove Arup and Partners (no date). Desert sand problems and their control, unpublished report to the Saudi Arabian Ministry of Communications, Technical Administration for Roads and Ports, 97 pp.

Petrov, V. I. (1983). Mobile sand fixation in the arid zone of the R.F.S.F.R., *Problems of Desert Development*, No. 5, 67–70.

Phillips, C. J. and Willetts, B. B. (1979). Predicting sand deposition at porous fences, *Journal of Waterway, Port, Coastal and Ocean Division*, **14379**, 15–31.

Polyakova, Y. Y. (1976). Polymers—soil conditioners and nitrogen fertilizers, *Soviet Soil Science*, **8**, 443–6.

Prishchepa, A. V. (1984). The wind erosion status of sandy surfaces in the zone of phase I of the Kara Kum Canal, *Problems of Desert Development*, No. 2, 79–83.

Pye, K. (1980). Beach salcrete and eolian sand transport: evidence from North Queensland, *Journal of Sedimentary Petrology*, **50**, 257–61.

Ritchie, W. (1974). Environmental problems associated with a pipeline landfall in coastal dunes at Cruden Bay, Aberdeenshire, Scotland, *Proceedings of the Coastal Engineering Conference (Copenhagen)*, 1974, pp. 2568–80.

Rubin, D. M. and Hunter, R. E. (1985). Why deposits of longitudinal dunes are rarely recognized in the geological record, *Sedimentology*, **32**, 147–57.

Sarnthein, M. and Walger, E. (1974). Die äolische Sandstrom aus der W-Sahara zur Atlantikküste, *Geologische Rundschau*, **63**, 1065–87.

Sarre, R. D. (1988). Evaluation of aeolian sand transport equations using intertidal zone measurements, Saunton Sands, England, *Sedimentology*, **35**, 671–9.

Savage, R. P. and Woodhouse, W. W. (1968). Creation and stabilization of coastal barrier dunes, *Proceedings of the Coastal Engineering Conference (London)*, No. 1. pp. 671–700.

Schyfsma, E. (1978). Climate, in *Quaternary Period in Saudi Arabia* (Eds. S. S. Al-Sayari and J. G. Zötl), Springer-Verlag, Wien.

Shirmamedov, M. (1978). Protection of railways against sand drift in western Turkmenistan, *Problems of Desert Development*, **1978**, 86–9.

Simons, F. S. (1956). A note on the Pur-Pur dune, Virú Valley, Peru, *Journal of Geology*, **64**, 517–21.

Smith, R. S. U. (1981). Birth and death of barchan dunes in the southern Algodones dune chain, California and Mexico, *Geological Society of America, Abstracts with Programs*, **13**, 107.

Snyder, M. R. and Pinet, P. R. (1980). Morphological differences of coastal dunes created with straight and zig-zag sand fencing: implications regarding erosion along Westhampton Beach, Long Island, *Geological Society of America, Abstracts with Programs*, **12**, 84.

Stevens, J. H. (1974). Sand stabilization in Saudi Arabia's Al-Hasa Oasis, *Journal of Soil and Water Conservation*, **29**, 129–33.

Svasek, J. N. and Terwindt, J. H. J. (1974). Measurements of sand transport by wind on a natural beach, *Sedimentology*, **21**, 311–22.

Talbot, M. R. (1984). Late Pleistocene rainfall and dune building in the Sahel, *Palaeoecology of Africa and the Surrounding Islands*, **16**, 203–14.

Thomas, D. S. G. (1987). Discrimination of depositional environments using sedimentary characteristics in the Mega Kalahari, central southern Africa, in *Desert Sediments: Ancient and Modern* (Eds. L. Frostick and I. Reid), pp. 293–306, Geological Society (London), Special Publication 35.

Thomas, D. S. G. (1988). Analysis of linear dune sediment-form relationships in the Kalahari dune desert, *Earth Surface Processes and Landforms*, **13**, 545–53.

Thomas, D. S. G. and Goudie, A. S. (1984). Ancient ergs of the southern hemisphere, in *Late Cainozoic Palaeoclimates of the Southern Hemisphere* (Ed. J. C. Vogel), pp. 407–18, Balkema, Rotterdam.

Trossel, C. J. (1981). Eolian sand control in Saudi Arabia as experienced by ARAMCO, unpublished paper presented at the Symposium on Geotechnical Problems in Saudi Arabia, Riyadh, 1981.

Tsoar, H. (1982). Internal structure and surface geometry of longitudinal (seif) dunes, *Journal of Sedimentary Petrology*, **52**, 823–31.

Tsoar, H. (1983). Wind tunnel modelling of echo and climbing dunes, in *Eolian Sediments and Processes* (Eds. M. E. Brookfield and T. S. Ahlbrandt), pp. 247–59, Elsevier, Amsterdam.

Tsoar, H. (1984). The formation of seif dunes from barchans—a discussion, *Zeitschrift für Geomorphologie, N.F.*, **28**, 99–103.

Tsoar, H. (1985). Profiles analysis of sand dunes and their steady state signification, *Geografiska Annaler*, **67A**, 47–59.

Tsoar, H. (1986). The advance mechanism of longitudinal dunes, in *Physics of Desertification* (Eds. F. El-Baz and M. H. A. Hassan), pp. 241–50, Martinus Nijhoff, Dordecht.

Tsoar, H. and Møller, J. T. (1986). The role of vegetation in the formation of linear sand dunes, in *Aeolian Geomorphology* (Ed. W. G. Nickling), pp. 75–95, Allen and Unwin, Boston, Mass.

Tsoar, H. and Zohar, Y. (1985). Desert dune sand and its potential for modern agricultural development, in *Desert Development* (Ed. Y. Gradus), pp. 184–200, D. Reidel, Dordecht.

Twidale, C. R. (1981). Age and origin of longitudinal dunes in the Simpson and other sand ridge deserts, *Die Erde*, **112**, 231–47.

Ungar, J. E. and Haff, P. K. (1987). Steady state saltation in air, *Sedimentology*, **34**, 289–99.

USDA, Soil Conservation Service (no date). Building, planting and maintaining costal sand dunes, Conservation Information 32.

USDA, Soil Conservation Service (1977). 'Cape' American beachgrass: conservation plant for mid-Atlantic sand dunes, Program Aid 1152.

USDA, Soil Conservation Service (1982). 'Atlantic' coastal panicgrass, Program Aid 1318.

Verstappen, H. T. (1968). On the origin of longitudinal (seif) dunes, *Zeitschrift für Geomorphologie, N.F.*, **12**, 200–20.

Vinogradov, B. V. and Kulik, K. N. (1987). Aerospace monitoring of desertification dynamics of black earth lands of Kalmykia according to sequential photos, *Problems of Desert Development*, No. 4, 47–55.

Warren, A. (1972). Observations on dunes and bi-modal sands in the Ténéré desert, *Sedimentology*, **19**, 37–44.

Warren, A. and Knott, P. (1983). Desert dunes: a short review of needs in desert dune research and a recent study of micrometeorological dune-initiation mechanisms, in *Eolian Sediments and Processes* (Eds. M. E. Brookfield and T. S. Ahlbrandt), pp. 343–52, Elsevier, Amsterdam.

Wasson, R. J. and Hyde, R. (1983). A test of granulometric control of desert dune geometry, *Earth Surface Processes and Landforms*, **8**, 301–12.

Watson, A. (1985). The control of wind blown sand and moving dunes: a review of methods of sand control in deserts, with observations from Saudi Arabia, *Quarterly Journal of Engineering Geology, London*, **18**, 237–52.

Watson, A. (1986). Grain-size variations on a longitudinal dune and a barchan dune, *Sedimentary Geology*, **46**, 49–66.

Watson, A. (1987). Discussion of 'Variations in wind velocity and sand transport on the windward flanks of desert sand dunes', *Sedimentology*, **34**, 511–20.

Webb, R. H., Steiger, J. W. and Newman, E. B. (1988). The response of vegetation to disturbance in Death Valley National Monument, California, United States Geological Survey Bulletin, 1793, 103 pp.

Whitney, M. I. (1978). The role of vorticity in developing lineation by wind erosion, *Bulletin of the Geological Society of America*, **89**, 1–18.

Whitney, M. I. (1983). Eolian features shaped by aerodynamic and vorticity processes, in *Eolian Sediments and Processes* (Eds. M. E. Brookfield and T. S. Ahlbrandt), pp. 223–45, Elsevier, Amsterdam.

Willetts, B. B. (1983). Transport by wind of granular materials of different grain shapes and densities, *Sedimentology*, **30**, 669–79.

Willetts, B. B. and Phillips, C. J. (1978). Using fences to create and stabilize sand dunes, *Proceedings of the Coastal Engineering Conference (USA)*, No. 2, 2040–50.

Wilson, I. G. (1972). Aeolian bedforms—their development and origins, *Sedimentology*, **19**, 173–210.

Wilson, I. G. (1973). Ergs, *Sedimentary Geology*, **10**, 77–106.

CHAPTER 3

Wind Erosion and Dust-storm Control

N. J. MIDDLETON

INTRODUCTION

The movement of soil particles by wind occurs in many environments, but it is most pronounced and causes the most serious problems in the world's dry lands. The normally sparse vegetation cover and low, erratic rainfall characteristic of dry areas predisposes them to the importance of wind as a major erosional agent, but increasing human pressures in the world's semi-arid and arid lands has meant that accelerated soil erosion by wind is closely linked to the processes and patterns of land degradation and desertification.

Attempts to control wind erosion and dust-storm occurrence and ameliorate the problems they cause must be based on a thorough knowledge of the nature of aeolian soil movement. This chapter will examine the mechanics of wind erosion; identify the areas where it has been and is most prevalent and the actions that accelerate the process; outline the consequences of wind erosion; and review the methods available for its control.

NATURE OF WIND EROSION

When and where wind erosion occurs is determined by the mutual interaction between the elements of wind erosivity and surface erodibility (Figure 3.1). The movement of particles, indicating a departure from the stable condition, may be initiated by a change, positive or negative, in one or more of the variables shown in Figure 3.1. In the field these variables of erosivity and erodibility change through time and space, at varying rates and differing scales, so that the relationship between variables is in a constant state of flux. In order to ease description, it is appropriate firstly to look at the role of wind,

Techniques for Desert Reclamation
Edited by A. S. Goudie
© 1990 John Wiley & Sons Ltd.

EROSIVITY

WIND VARIABLES
Velocity	−
Frequency	−
Duration	−
Magnitude	−
Shear	−
Turbulence	−

ERODIBILITY

DEBRIS VARIABLES
Particle size	±
Soil clods and cohesive properties	+
Abradability	−
Transportability	−
Organic matter	+

SURFACE VARIABLES
Vegetation :	residue	+
	height	+
	orientation	+
	density	+
	fineness	+
	cover	+
Soil and moisture		+
Surface roughness		+
Surface length (distance from shelter)		−
Surface slope		±

Figure 3.1   Key variables in the wind erosion system. Wind erosion will normally be reduced if the values of variables are increased (+) and if other variables are reduced (−) (modified after Cooke and Doornkamp, 1974)

followed by that of soil, the threshold velocities of soil entrainment, and the effects of ground surface characteristics.

Wind

Wind moves soil by virtue of its energy, and the availability of a wind's energy to promote entrainment is related to a large number of atmospheric variables in an extremely complicated manner. As such, the details concerning wind velocity, turbulence, gustiness, shear forces, humidity, and temperature, and their relationships in the wind erosion system are not well understood, but some comments can be made on work done in relation to wind characteristics to date.

Perhaps the most important of these wind variables is velocity, although the limitations of this measure have been indicated by Wilson and Cooke (1980) who quote a number of measures of turbulence to illustrate the variety of conditions that may be experienced given the same mean wind velocity. Wind tunnel and field measurements have shown that the rate of soil movement is proportional to the third power of mean wind velocity, and Skidmore and Woodruff (1968) applied this relationship in their assessment of wind erosion forces for 212 locations in the USA. For each location three measures were calculated: the magnitude of wind erosion forces, the prevailing wind direction, and the preponderance of wind erosion forces in the prevailing wind erosion direction. The magnitude is calculated for each month in each of sixteen compass points and the total magnitude for each month is given by summing the magnitude from each direction, giving a measure of the relative capacity of the wind to cause soil erosion. The prevailing wind erosion direction is based upon the ratio of forces parallel and perpendicular to the prevailing direction, its final value being a measure of the preponderance of the prevailing direction.

A major deficiency of this technique is that the wind velocity measures include all winds, irrespective of whether or not they exceed the threshold velocity for particle movement at a particular location. This problem is overcome in an index developed by Fryberger (1979) to express wind forces in terms of their potential for sand transport. The index is based on Lettau and Lettau's (1978) equation for the rate of sand drift, from which an expression is derived for the annual or monthly rate of sand drift:

$$Q \propto V^2(V - V_t)t$$

where Q = annual or monthly rate of sand drift
V = wind velocity at 10 m height
V_t − impact threshold wind velocity at 10 m height
t = length of time wind blew, expressed as a percentage of a year's observations

Fryberger uses an impact threshold velocity of 12 knots. The results are expressed in vector units and the vector unit total for any wind summary is proportional to the sand-moving power of the wind at the station of record, and is known as the drift potential (DP). The applicability of the DP as an appropriate measure of soil erosion potential depends on threshold velocities for soil entrainment, but Fryberger's equation has been used with success as an indicator of dust-raising potential in West Africa (Middleton, 1987) and of potential soil erosion in Australia (Kalma *et al.*, 1988). In a number of works by Gillette and his coworkers, field studies have been undertaken of thresholds, and these are reviewed below.

A deficiency common to both Skidmore and Woodruff's and Fryberger's indices is that the measures fail to take account of all weather conditions that might be associated with winds from a particular direction. Thus, for example, the prevalent wind might typically bring rainfall while most of the erosion is caused by less frequent but drier winds from another direction.

Many studies show that soil drifting increases with length of exposed area. A wind may start to entrain material at the windward end of a field and continue to increase its load until it can carry no more if the exposed area is long enough. As a wind continues it may pick up material but it will also deposit some of its load since its carrying capacity is finite. The maximum rate of transport of a wind with a specific friction velocity is very similar for all soils, although the exposed length needed for a wind to reach load capacity depends upon the erodibility of the particular soil. The more erodible the soil, the shorter the distance to reach carrying capacity. The distance required for the maximum load to be picked up by a wind of $18 \, \text{ms}^{-1}$ at 10 m above the ground varies from less than 55 m for a structureless fine sand to more than 1500 m for a cloddy medium-textured soil (Chepil and Woodruff, 1963).

Soil

Soil erodibility depends largely on the mechanical stability of the soil, which is defined by Chepil and Woodruff (1963) as the resistance of a dry soil to breakdown by a mechanical agent such as tillage, force of wind, or abrasion from wind-blown materials. This mechanical stability is dependent upon the size, density, and shape of its individual particles, and most soils consist of individual particles held together by various forces as aggregates or clods of varying sizes. Smalley (1970) points out that the mechanical stability of a soil depends largely on these forces of interparticulate cohesion in the soil system. Perhaps the most important binding agents for these soil structures are the ratio of sand, silt, and clay particles within them, their soil moisture properties, and the presence of cements such as salts and those associated with decomposing organic matter.

Chepil and Woodruff (1963) have investigated the relative effectiveness of silt and clay as binding agents, which depends somewhat on their relative proportions to each other and to the sand fraction. It is only dry soil particles that are readily erodible by wind since soil moisture promotes particle cohesion and thus restricts erodibility (Bisal and Hsieh, 1966). Also, soils differing in texture require different percentages of moisture to resist initiation of soil movement at specific wind velocities. Experimental work by Bisal and Hsieh confirms the general rule first suggested by Chepil (1955a) that the higher the proportion of silt and clay in a soil, the greater the production of clods and the lower the erodibility. The moisture content of a soil at any particular time is in turn determined by the properties of that soil and by particular weather conditions.

Water also affects the wind erosion system in the form of raindrop impact, which may form surface crusts, normally consisting of silt and clay sized particles, with coarser particles left loose on the surface (Chepil and Wood-ruff, 1963). These loose particles are easily dried and may be moved by the wind soon after rainfall has ceased, contributing to the initial stages of wind erosion, breaking down the crust by abrasion and enhancing further drying. Rainfall may also reduce erosion through its effect on plant growth (see below).

A variety of cements that decrease a soil's erodibility can be identified. One of the commonest in the desert landscape is salt which may act to combine particles and is also present in the form of hard crusts that cover large areas of the desert surface.

A number of cements is produced from the breakdown of organic materials by microorganisms. These cements are derived from the decomposition products of plant residues, the decomposer microorganisms themselves, and their secretory products. Together they serve to bind particles together thus improving soil structure. Chepil (1955b) found that between 1 and 6% organic matter added to a soil in the early stages of decomposition (less than one year) led to enhanced clod production and decreased erodibility. Over a period of four years, however, there was a decline in clod production and consequent increase in erodibility, as initial cementing materials change, lose their cementing properties, and become brittle when microbial activities diminish. The microbial fibres also disintegrate in time and a high proportion of medium-sized water-stable aggregates develop which are highly susceptible to wind erosion (Troeh *et al.*, 1980). This point is particularly relevant in areas susceptible to prolonged drought, where several years of below-average rainfall will indirectly result in less organic matter reaching a soil surface through its effects on vegetation growth.

Threshold Velocities

In recent years Gillette and his coworkers have attempted to determine threshold erosion velocities for a variety of natural desert soil types. Investigation by Gillette *et al.* (1980) in the Mojave Desert found that the size distributions of saltating grains were practically the same as size distributions of the loose aggregates of the surface from which the saltating grains were generated. A definite trend of increasing threshold velocity with larger mode of the aggregate size distribution of soils was found, threshold velocity increasing approximately as the half-power of the size of the mode of the aggregate size distribution. The data compared favourably to Chepil's (1951) data for threshold velocity versus particle size in laboratory wind tunnel tests of simplified soils in which particles are all the same size. Thus, if the mode of the mass–size distribution of the loose particles present on a surface is known, it can be used as an approximate predictor for the threshold velocity of movement for that soil.

Ground Surface Characteristics

Since threshold wind erosion velocities are determined by the availability of loose particles at the soil surface, the threshold velocity will be increased by the effect of non-erodible elements, such as pebbles and larger objects and vegetation. Wind erosion will continue on a surface until a sufficient number of non-erodible elements are uncovered to provide cover and shelter to remaining erodible grains, although this situation may alter should the wind change direction or velocity.

The effect of non-erodible elements is most obvious on so-called 'wind-stable' surfaces such as stone pavements. Stone pavements occur widely in environments with little vegetation such as hot deserts, and generally act to protect otherwise potentially wind-erodible surfaces such as residual weathering mantles or alluvium (Cooke, 1970). Deflation can occur from desert pavements, however (Chepil and Woodruff, 1963), and it may be that the wind-stable surface, in reducing the high-frequency/low-intensity deflation events, allows the build-up of a reservoir of fine material that is susceptible to removal during a violent wind storm, as Bagnold (1941) suggests.

Surface ridges, produced by tillage, affect the quantity of a soil that is eroded. This effect is dependent upon the height and lateral frequency of ridges, their shape, orientation to the wind direction, and their proportion of erodible to non-erodible grains. The most effective orientation of ridges is at right angles to the erosive wind, and as the wind moves closer to being parallel to them their effect decreases to the point where they may actually enhance soil loss by the encouragement of a scouring effect along the furrows.

Vegetation influences the nature of wind erosion in several ways. The quantity (proportion of ground surface covered) and quality (height, density, and flexibility) of vegetation governs the extent to which a surface is exposed to erosion and the degree by which surface roughness is increased. These properties will of course vary with vegetation type and, for a given type, according to the season. In general, the taller the crop, the finer the vegetative material, and the greater its surface area the more the wind velocity is reduced. Chepil and Woodruff (1963) suggest that grass offers one of the most effective protective covers.

Vegetation also stabilizes soil structures through its root systems, and vegetative decay adds organic matter to the soil. Plant litter is important in protecting the soil surface, as it both reduces wind velocity and traps eroded material, as well as by contributing organic cements. With this understanding of the way in which vegetation effects the erodibility of soil it follows that the removal of vegetation will increase the risk of wind erosion and dust-storm generation. Such removal may occur naturally during drought periods and in many instances human and animal populations may also alter and destroy vegetation cover.

Topography affects the wind erosion system in a number of ways. Over relatively short slopes wind shear is greater at the upper part of the windward slope (Troeh *et al.*, 1980), and the presence of knolls or hollows, for example, is likely to affect such variables of erodibility as soil moisture (Wilson and Cooke, 1980).

Larger-scale features such as valleys can locally enhance wind velocities by the 'venturi effect', to more easily entrain fine sediments that are usually transported to valley bottoms. Topography also plays a part in the formation of particular meteorological systems that produce strong winds capable of erosion. Thus, the intense solar heating of wide flat desert landscapes during daylight hours induces convective activity and the production of turbulent flow, and mountain and valley slopes may induce katabatic flow that locally increases wind speeds.

REGIONS OF WIND EROSION

Several authors have provided a global view of the most active wind erosion regions using standard meteorological data of dust-storm occurrence, with some corroboration from remote-sensing platforms (e.g. Grigoryev and Kondratyev, 1980; Goudie, 1983; Middleton *et al.*, 1986). Broadly, the areas of most intense activity occur in parts of the Sahara Desert, the Middle East, Soviet Central Asia, and northern China/Mongolia. In North Africa and the Middle East the most active areas as determined by the above authors and more detailed regional appraisals (Middleton, 1986a, 1986b) are generally in good correspondence with the regions at risk from wind erosion shown on FAO/UNEP/UNESCO (1980) maps of soil degradation risks. The methodology used for the preparation of these maps include parameters of climate, soil, topography, and human action (FAO, 1979).

Using dust-storm and rainfall data for a number of world regions, Goudie (1983) has shown that deflation of dust is at a maximum in areas with an annual rainfall from 100 to 200 mm. This is in contrast to areas with <100 mm annual rainfall, where dust-storm frequency declines. A summary of the geomorphological terrain types from which substantial deflation occurs shows that most are fluvial or lacustrine in origin (Middleton, 1989) so that infrequent stream runoff and thus limited dust supply may explain Goudie's observations. Goudie also suggests that strong winds associated with fronts and cyclonic disturbances are rare in hyper arid regions. Although fluvial processes are more active in desert marginal areas, therefore providing greater supplies of fine sediments for deflation, it is also possible that recent cultivation of desert marginal soils is a more important factor behind the peak of dust-storm activity in the 100–200 mm annual rainfall zone (Pye, 1987). A number of case studies where the human impact has been a significant factor in instances of wind erosion will now be reviewed in order to shed more light on the anthropogenic input to the wind erosion system.

Maghreb

A significant turning point in the course of environmental degradation in the Maghreb countries of North Africa is linked to the change in cultivation practices brought to the region by the French settlers in the nineteenth century (Dresch, 1986). Before the French arrived the traditional cultivation tool of the dry land 'telle' zone of Algeria and Tunisia was the hoe. This hand-held tool 'scratched' rather than overturned the soil, but did so to a depth enough to allow grain shoots through. Such cultivation was carried out every other year, while during intervening years the land was left fallow and sheep turned out to graze.

The French colonists brought to North Africa the heavy farming implements used in nineteenth century France:

'The Brabant plough, complete with the forecarriage, share, furrow-opener, and mouldboard, turned the layer to twice the depth the hoe did; seed scarification, harrowing, etc., that followed ploughing, crushed the clods, thereby changing altogether the soil structure, and removed shrubs and weeds from the field leaving it "clean"' (Dresch, 1986, p. 67).

Further destruction of soil structure was perpetrated by the adoption of American-made rain-fed farming implements such as the multishare and multidisc ploughs, scarifiers, and harrows.

The expansion of colonialists onto the traditional croplands pushed indigenous farmers to plough up hillsides and mountain slopes formerly used as rangelands. By the early twentieth century the cropland area had quadrupled while the dry cereal crops had advanced to their climatic limits of about 250 mm of annual rainfall. Herdsmen whose seasonal grazing had been put to the plough were pushed further south. With independence the situation has not changed, with the desire to enlarge cropland areas pushing the tractor and multidisc plough deeper into the steppe and more fragile ecosystems. The result, as Dresch sees it, has been increasing sensitivity to wind erosion and a general degradation of the steppe ecosystem.

The Great Plains of the USA

The Great Plains is an area prone to drought and accompanying windstorms, where human actions have at times had major impacts on the wind erosion system. The Dust Bowl of the 1930s is perhaps the best-known and most oft-quoted example of large-scale wind erosion and dust-storm activity anywhere in the world. The core of the dust bowl area comprised the western third of Kansas, south-east Colorado, the Oklahoma Panhandle, the northern two-thirds of the Texas Panhandle, and north-east New Mexico, although most of the Great Plains experienced dust bowl conditions at some time during the 1930s. Indeed some of the worst conditions were found as far north as Wyoming, Nebraska, and the Dakotas.

The most severe dust-storms occurred in the Dust Bowl between 1933 and 1938, with activity being at a maximum during the spring of these years. At Amarillo, Texas, at the height of the period, one month had 23 days with at least ten hours of air-borne dust, and one in five storms had zero visibility (Choun, 1936). For comparison, the long-term average for this part of Texas is just six dust-storms a year (Changery, 1983).

The reasons for this most dramatic of ecological disasters have been widely discussed, and blame has largely been laid at the feet of the pioneering farmers and 'sod busters' who ploughed up the plains for cultivation. For although dust-storms are frequent in the area during dry years, and the 1930s was a period of drought, the scale and extent of the 1930s events were unprecedented.

Cultivation of the Great Plains started in the late 1870s, and the natural sod-forming grasslands were slowly transformed into wheat fields. The waves of settlers that arrived in the area from 1914 to 1930, in conjunction with the increasing use of mechanized agriculture, catalysed by high wheat prices, led to unprecedented large-scale wind erosion when drought hit the plains in 1931. In 1937 the US Soil Conservation Service estimated that 43% of a 6.5 million hectare area in the heart of the Dust Bowl had been seriously damaged by wind erosion. One event, that raised material in Montana and Wyoming on 9 May 1934, carried an estimated 350 million tons of soil eastward, depositing dust in Boston and New York in the morning of 11 May and on ships' decks 500 km out in the Atlantic during the next day or so (Worster, 1979).

There is an approximate twenty-year drought cycle in the Great Plains. Major droughts have occurred in the 1890s, 1910s, 1930s, 1950s and 1970s, and these droughts are normally periods of exaggerated dust-storm activity. Although the dust-storms of the 1950s were not as spectacular as those of the 1930s, more land was actually damaged by wind erosion in the Great Plains, and soil loss in the 1970s was on a scale comparable to that of the 1930s (Lockeretz, 1978).

In the 1970s a new phase of inappropriate cultivation practices was highlighted. Perhaps the worst single dust storm occurred after two years of drought, in the Portales Valley area of eastern New Mexico, as a low-pressure frontal system moved eastwards across the Great Plains on 23 February 1977. In the Portales Valley dryland wheat farming had moved onto marginal land as a result of economic factors and high-technology land-use practices. In addition, certain government policies reduced the disincentives to cultivate marginal land, so that the Wheat Disaster Assistance Program compensated farmers for loss of crops to wind erosion. Thus, cultivation of wheat on marginal lands, some of which were formerly dune fields and wind-deposited loess, was encouraged by high prices that followed export sales of great quantities of wheat to the USSR in 1975. These crops were irrigated by centre-pivot irrigation which requires the removal of linear wind breaks made up of trees planted since the Dust Bowl era to help prevent wind erosion (McCauley *et al.*, 1981). Some farmers ploughed their fields parallel to rather than transverse to the erosive wind direction, and some had failed to leave a protective cover of stubble mulch on their fields. The dust palls from Portales Valley and an area in eastern Colorado/western Kansas were tracked on SMS-2 and GOES-1 satellite imagery south-eastwards, obscuring 400 000 km^2 of ground surface in south-central USA and out over the mid-Atlantic Ocean (Breed and McCauley, 1986).

The Virgin Lands

The costly ecological lessons of converting grasslands to cereal production without due regard for the environmental conditions have also been learnt the hard way in the USSR. When Khrushchev came to power in 1953 he inherited an agricultural system that was barely producing as much food as in prerevolutionary days forty years before. The answer to the country's need to massively increase grain production was to plough up the 'Virgin Lands' of northern Kazakhstan, western Siberia, and eastern Russia. Between 1954 and 1960 forty million hectares of new land were brought under cultivation and this new land was mainly responsible for a 50% increase in grain output over the same period (Eckholm, 1976).

Deep ploughing was used, which removed the stubble from previous crops, allowing planting earlier in the year, thus reducing the chance of widespread crop loss to an early snow at harvest time. The land was also used more intensively than by traditional practices, which left millions of hectares fallow under grass each year. The alternative 'crop fallow' system recommended corn to be planted in what were supposed to be rest years between wheat plantings.

The dangers of these methods in a drought-prone region were becoming clear soon after the Virgin Lands campaign started. In one Kazakhstan Oblast alone Chakvetadze (1962) noted that more than one million hectares of crops were damaged by wind erosion in the period 1955–60. However, the devastation reached a peak in the drought year of 1963, when crops on three million hectares of Virgin Land were lost altogether to the drought and 'the normally savage winds carried precious topsoil, now dehydrated and easily torn from the earth, off the farms' (Eckholm, 1976, p. 56).

The Sahel

The Sahel region on the southern margins of the Sahara Desert in North Africa is the most recent world region to be subjected to increased wind erosion during a period of drought and when human degradation of the natural vegetation has been widely reported.

During the severe drought years of 1972–4 mean concentrations of Sahelian soil dust blown to Barbados were three times their predrought levels (Prospero and Nees, 1977) and similarly high concentrations were also recorded in the early 1980s (Prospero and Nees, 1986). In the Sahel itself Bertrand *et al.* (1979) showed increasing trends in days with dust haze that rose dramatically during the below-average rainfall period that began in 1968–9 at Niamey, Zinder, and Agadez in Niger. To the east in Sudan and to the west in Mauritania, Middleton (1985, 1987) has shown increasing dust-storms to be related to decreasing rainfall during the 1970s and 1980s.

Although the 1970s and 1980s have been periods of prolonged drought in parts of the Sahel, these and previous decades have also been highlighted as times of anthropogenic ecological degradation in the region (e.g. Rapp, 1974; UN, 1977; UNEP, 1985). Among the commonly quoted human activities that are responsible partly or wholly for desertification are overgrazing, overcultivation, woodcutting for fuel and agricultural clearance, salinization and other forms of vegetation destruction, and surface destabilization from an expanding population. However, although the theory relating such activities to increased wind erosion is sound, there are, to date, few specific studies proving human action to be instrumental in enhanced soil deflation in the Sahel.

EFFECTS OF WIND EROSION

The movement of soil and other sediments by wind has a large number of environmental impacts beyond the significant consequences for farmers. The problems associated with wind erosion can be classified and examined according to the three fundamental processes: deflation, transport, and deposition.

Problems of Deflation

Deflation of soils removes the finest particles, which are some of the most important soil constituents: clay, silt, and organic matter. The ratio of sand, silt, and clay sized particles is of primary importance to a soil's stability (Chepil and Woodruff, 1963) and thus the preferential removal of silt and clay is detrimental to soil structure. These particles also have important influences on a soil's moisture-retention capacity and, thus, their removal also reduces soil moisture storage. The maximum concentration of a soil's nutrients is also attached to the finest particles (e.g. Gupta *et al.*, 1981) and thus removal of the fines reduces fertility. Estimates of organic matter and nutrient loss during dust-storms in Rajasthan have been made by Wasson and Nanninga (1986) who calculate rates of 124 T ha for organic matter and 100 T ha for N, P, and K nutrients.

Studies in the Great Plains by Fryrear (1981) show long-term decreasing trends in yields of dryland crops in the Texas Panhandle. Thirty to forty consecutive years of yield data showed that sorghum yields had declined by 67% and kafir yields by 59%. Water erosion, annual cropping practices, and increasing insect and disease hazards were suggested in addition to wind erosion as possible causes. Fryrear notes that improvements in crop varieties and cultivation practices should have partially compensated for decreasing yields, but had obviously not kept pace with the factors responsible for decreased crop production. Surface soil in drylands does not regenerate quickly under natural conditions, so that without changes in cultivation practices deflation loss may be fairly permanent.

Other effects of deflation include the scouring and undermining of structures such as telegraph poles, fencing, railway sleepers, and roads. Such effects may lead to the collapse of structures (Cooke *et al.*, 1982).

Problems during Transport

The transport of soil particles near the ground surface can present serious problems of abrasion. Soil clods may be disintegrated as a result of saltating grain impact, thus impoverishing soil structure and rendering soil more erodible (Chepil, 1946). Indeed, soil erosion of a field initiated from a small, highly erosive spot, from which soil particles may be removed by direct wind pressure, is often sufficient to initiate erosion over an entire field as saltating grains start movement of other grains by transfer of their energy in impact (Chepil, 1946).

Crops can also be abraded and in extreme cases cut from their exposed roots by saltating grains, and a layer of soil dust decreases the marketability of vegetable crops such as asparagus, green beans, and lettuce (Skidmore, 1986).

The abrasion of structures is also important. Sand abrasion can have significant effects on building structures, producing pitting, flutes, and grooves. In the Seistan Basin of south-west Afghanistan the potential of saltating grains to frost the glass in windows has meant that houses are built with dead walls facing the prevailing 'Wind of 120 days' (Middleton, 1986b). Paintwork can easily suffer, telegraph poles and fences blasted at their base.

Large-scale atmospheric dust concentrations affect local meteorological processes and may over long periods lead to reduced rainfall (Bryson and Baerreis, 1967; Wells and Middleton, 1988). Dust-sized material transported in suspension can seriously affect transport facilities through visibility reduction. Airports may have to be closed during severe events (e.g. Houseman, 1961) and poor visibility has caused a number of aircraft accidents (Pye, 1987). On the roads the dangers of sudden visibility reduction in Arizona has inspired the development of a Dust Storm Alert System incorporating remotely controlled road signs and dust alert messages on local radio (Burritt and Hyers, 1981).

Inhalation of fine particles can aggravate human diseases such as bronchitis and emphesyma, and the transport of soil pathogens spreads human and plant diseases (e.g. Leathers, 1981; Clafin *et al.*, 1973). The suffocation of animals was reported during severe dust-storms during the Dust Bowl (Choun, 1936). Radio and satellite communications are adversely affected by atmospheric soil dust, solar power potential is decreased, and mechanical equipment affected.

Problems of Deposition

The deposition of drifting soil and soil dust can bury and kill plants, fill ditches, and block roads. The infiltration of dust into homes creates problems of sanitation and housekeeping and contamination of food and drinking water. Salt transferred in aeolian dust can be highly destructive of building materials and increase the salinity of groundwater. Clements *et al.* (1963) report that dust between relay contacts and abrasion of switches causes particular problems for a telephone company in California.

Deposition of sand-sized particles in sand drifts or moving dunes can completely bury urban obstacles in their path, including roads, railways, runways, pipelines, and cultivated gardens. Burial of pipelines and similar features poses problems for maintenance and inspection and may exert strains sufficient to cause fracture (Cooke *et al.*, 1982).

There are some positive aspects to aeolian deposition that are worth highlighting. The high nutrient content of dust particles provides additional fertility where they are deposited on terrestrial or marine ecosystems. The loess plateau of northern China, for example, is the country's most productive wheat-growing region. Deposition of air-borne nutrients on the leaves of certain plants may be beneficial. Das (1988) suggests that dust is a significant source of nutrients on the leaves of some graminaceous plants such as rice, wheat, and grasses.

WIND EROSION CONTROL AND SOIL CONSERVATION

From the above discussion on the mechanics of wind entrainment and transport of soil particles it follows that control strategies must focus on improving the aggregate stability of a soil and increasing surface roughness to reduce wind velocity. The conservation methods available to field workers can be analysed under the widely accepted headings of agronomic measures, soil management, and mechanical methods (Stallings, 1957; Troeh *et al.*, 1980; Morgan, 1986). Agronomic or biological measures employ the role of vegetation to minimize erosion by protection. Soil management methods focus on ways of preparing the soil to promote good vegetative growth and improve soil structure in order to increase resistance to erosion. Mechanical methods manipulate the surface topography in ways such as installing shelter-belts or creating ridges so as to decrease the velocity of airflow.

The wind erosion equation (Woodruff and Siddoway, 1965), developed as a result of many investigations into factors controlling wind erosion in the Great Plains of the USA, remains a useful guide to the principles and elements of wind erosion control. The fundamental functional relationship is expressed as

$$E = f(I, K, C, L, V)$$

where E = potential average annual soil loss per unit area
I = soil erodibility index
K = soil ridge roughness factor
C = a climatic factor
L = unsheltered median travel distance of wind across a field
V = the equivalent density of vegetative cover

The elements in this wind erosion equation represent those that can be the target of control efforts. The identification of the areas that should be the targets for control may be straightforward where the problems are those of entrainment. Efforts to control problems of transport and deposition, however, are dependent upon accurate identification of source areas. The study of dust-storms on the global scale, for example, still needs to pinpoint specific source areas for many of the major dust-producing regions, although some progress has been made with regard to the types of ground surface that are susceptible to large-scale dust production (see above and, for example, Yaalon, 1987).

At the local scale, a study of meteorological records is necessary to identify prevailing wind erosion directions and wind erosivity, although adequate data are not always available. In their study of blowing sand and dust hazards in a Middle East settlement Jones *et al.* (1986) interpreted remote sensing images of geomorphology as proxy indicators of wind directions and demonstrated the need for thorough geomorphological mapping and field investigation of possible source areas. Such investigation can identify both present and potential future sources, thus allowing appropriate conservation measures to be employed and/or management strategies which may incorporate local legislation to control human activities and surface disturbance.

Agronomic Measures

Agronomic measures for controlling soil erosion utilize living vegetation or the residues from harvested crops to protect the soil. Standing crops and their residues protect soil by acting as non-erodible elements, absorbing the wind's shear stress. When a vegetative cover is sufficiently high and dense to prevent the wind stress on adjacent exposed land exceeding the threshold for particle movement, then the soil will not erode. Maintaining a sufficient vegetative cover is sometimes referred to as the 'cardinal rule' for controlling wind erosion (Skidmore, 1986).

The maintenance of crop residue or mulch as a stubble on cropland is recognized as an efficient method for reducing wind erosion losses. 'Stubble mulching' is a crop residue management technique that aims to maintain some degree of crop residue on the field surface at all times. The soil is usually tilled, but not to the extent that the field is left 'clean'. The tillage system usually utilizes blades or V-shaped sweeps and does not invert the soil (McCalla and Army, 1961).

Stubble mulching is a primary erosion control technique used in one method of 'conservation tillage' adopted by the Reagan administration in the USA in the mid 1980s. The farmer plants new seeds among the stalks and debris left from the previous harvest. The method reduces erosion and also reduces farmers' costs since fewer trips with tractor and ploughing equipment are needed through the fields.

There are drawbacks to leaving crop residues on fields, however. The residues often provide a good habitat for insects and weeds. In countries where pesticides are affordable this problem can be overcome with chemical applications, but with the concomitant hazards of off-field pollution, killings of non-target species, and development of resistance to the chemicals used. Where pesticides are not used, the insects and weeds will combine to reduce yields by eating crops and competing for soil nutrients.

Other forms of stabilizers made of synthetic materials have also been evaluated for their applicability to wind erosion control (e.g. Armbrust and Dickerson, 1971). Some stabilizers have been found to meet the essential criteria for soil surface stabilizers (Armbrust and Lyles, 1975).

(a) One hundred per cent of the soil must be covered.
(b) Stabilizers must not adversely affect plant growth or emergence.
(c) Erosion must be prevented initially and reduced throughout the period of severe erosion hazard.
(d) The stabilizer must be easily applied and without special equipment.
(e) The cost must be low enough for profitable use.

The practice of farming land in narrow strips, on which crop alternates with fallow usually of a leguminous or grass crop, is an effective wind erosion control technique. The most effective strips are perpendicular to the prevailing erosive wind direction, but they do provide some protection from winds not perpendicular to the field strip (Skidmore, 1986). The strips reduce the wind velocity across the fallow strip, reduce the distance the wind travels over exposed soil, and localizes any soil drifting. Strip cropping demands small fields, however, and thus is not compatible with highly mechanized agriculture, but provides a useful technique for the smallholder.

Soil Management

Soil management techniques essentially deal with different methods of tillage. Although tillage is an important management technique, providing a suitable seed bed for plant growth and helping to control weeds, the dangers of excessive tillage have been illustrated in the Maghreb, Great Plains, and Virgin Lands examples outlined above. Excessive tillage, particularly of light textured soils, breaks soil clods and exposes soil to wind action, particularly if soil overturning binds stubble into the soil thus reducing mulch coverage. To overcome this destruction of structure in non-cohesive soils, tillage operations must be restricted. This may be by reducing the number of passes over a field by combining as many operations into one pass as possible, such as in mulch tillage or minimum tillage, or by strip-zone tillage where operations are concentrated only as rows where the plant grows, leaving the inter-row areas untilled (Schwab *et al.*, 1966; see Table 3.1).

The effects of various forms of conservation tillage on erosion rates, soil conditions, and crop yields has been the subject of many studies in recent years and the results show the success of the system to be highly soil specific and also to depend on how well weeds, pests, and diseases are controlled (Morgan, 1986). The practice of no-tillage agriculture, in which drilling is carried out directly into the stubble of the previous crop, has been found to show great promise in recent years (e.g. Phillips *et al.*, 1980). It reduces labour costs, soil and moisture losses, and maintains good structure. Schmidt and Triplett (1967, quoted in Phillips *et al.*, 1980) showed soil erosion loss from a no-tillage field of corn in Ohio to be 4.5 T ha^{-1} during a severe windstorm as compared to a conventionally planted cornfield that lost 291 T ha^{-1}. In Nebraska the use of no-tillage and herbicides to control wind erosion resulted in less weed growth, higher soil moisture storage, and higher grain yields than conventional tillage over a six-year period (Wicks and Smith, 1973). No-tillage has been embraced by agrochemical companies because it requires heavier doses of pesticides, but this in itself is not necessarily desirable due to possible increases in off-field pollution (Risser, 1985). Plant residues on no-tillage fields may lower soil temperatures by as much as 6°C at 25 mm depth in spring, which can delay spring plantings in central and northern North America where soil temperatures are below those needed for optimal growth,

Table 3.1   Tillage practices used for soil conservation

Practice	Description
Conventional	Standard practice of ploughing with disc or mouldboard plough, one or more disc harrowings, a spike-tooth harrowing, and surface planting
Strip or zone tillage	Preparation of seed bed by conditioning the soil along narrow strips in and adjacent to the seed rows, leaving the intervening soil areas untilled: e.g. plough–plant; wheel-track planting; listing
Mulch tillage	Practice that leaves a large percentage of residual material (leaves, stalks, crowns, roots) on or near the surface as a protective mulch
Minimum tillage	Preparation of seed bed with minimal disturbance; use of chemicals to kill existing vegetation, followed by tillage to open only a narrow seedband to receive the seed; weed control by herbicides

After Schwab *et al.* (1966).

but in the tropics this effect may be useful where soil temperatures are frequently above the optimum for maximum plant growth (Phillips *et al.*, 1980). Nevertheless, experience in north-western India suggests that because of the low organic matter content of sandy soils in arid areas, they become compacted with no-tillage systems, which seriously reduces the growth and yield of crops (Gupta *et al.*, 1983).

A recent technique developed by the US Department of Agriculture in Arizona specifically for grassland revegetation involves the 'firming' and 'shaping' of the land surface. 'Land imprinting' refines the function of nature in which hoofprints from grazing ungulates perform the role of seed bed preparation by holding rainwater for soil infiltration and thereby allowing 'nature-irrigated' germination if a seed is present (Anderson, 1987). The imprinting machine consists of a single rolling cylinder, the only moving part, attached to a pulling frame. The imprints on the soil are made by angle irons welded to the cylinder; their configuration can be adapted to specific site conditions. The design is so simple that the machine can be made in any sophisticated welding workshop anywhere in the world.

Mechanical Methods

Mechanical approaches to wind erosion control include the creation of barriers to wind flow such as fences, windbreaks and shelter belts and altering surface topography such as by ploughing furrows.

Barriers to windflow aid erosion control by decreasing surface shear stress in their lee and by acting as a trap to moving particles, although barriers also create turbulence in their lee which can reduce their effective protection. Thus the most efficient fence, for example, is semi-permeable because, although its velocity reduction is less than for an impermeable fence, the amount of eddies and turbulence in its lee are reduced (Cooke *et al.*, 1982). In the same way vegetational windbreaks and shelterbelts should be designed to optimize the interaction between height, density, porosity, shape, and width of the vegetational barrier. A barrier oriented perpendicular to winds predominantly from a single direction will decrease wind erosion forces by more than 50% from the barrier leeward to 20 times its height, the decrease being greater at shorter distances from the barrier (Skidmore, 1986).

Windbreaks composed of a range of shrubs, tall-growing crops, and grasses, besides the more conventional tree windbreak, are used to control wind erosion. However, most barrier systems occupy space that would otherwise be used for crops. Perennial barriers grow slowly, can be difficult to establish and compete with the crops for water and plant nutrients (Dickerson *et al.*, 1976; Lyles *et al.*, 1983). Thus the net effects of tree barrier systems must be weighed against possible adverse effects on yields (e.g. Frank *et al.*, 1977).

The ploughing of ridges is a common antierosion measure which acts to roughen the soil surface and thus reduce the average wind velocity for some distance above the ground. Ridges also trap entrained particles on their leeward side (Chepil and Milne, 1941). Tillage to produce ridges across the path of the erosive wind is usually carried out by chisel and is successfully used temporarily to control wind erosion in an emergency (Woodruff *et al.*, 1957). Farmers of sandy soils in the Midlands counties of England employ a version of ridge and furrow tillage to control wind erosion on land devoted to sugar beet (Morgan, 1986). The Glassford system ploughs soil that is moist but not wet to produce ridges and furrows and immediately the furrows are rolled. The operation is carried out in January and the resulting furrowed and ridged surface remains stable throughout the spring blowing period.

CONCLUSION

The entrainment, transport, and deposition of soil surface particles by wind is a natural process common in the world's dry lands. The process presents a range of problems to inhabitants of such regions, problems that, with increasing agricultural and urban use of desert and desert-marginal lands, are becoming more important. This is not only because more people are occupying areas of wind erosion hazard, but also because human use of these areas all too often exacerbates the natural processes, making wind erosion more hazardous. The human actions most commonly responsible for this increase in wind erosion hazard are those that change or remove vegetation cover and those that destabilize natural surfaces. Such actions have a variety of motives: vegetation may be cleared for agriculture, building, fuel, or fodder; vegetation may be modified by cropping practices; land may be disturbed by ploughing, off-road vehicle use, military manoeuvres, construction, or trampling by animals. The offending actions are well known and documented and can be understood in the context of the considerable literature on the mechanics of soil, sand, and dust movement. There is also a range of control measures that result from considerable research, much of it carried out in the USA in response to the 'dirty thirties'. The amelioration of problems at specific sites must be based on investigation of factors of both erosivity and erodibility which, as Jones *et al.* (1986) suggest, should integrate remotely sensed data, meteorological data, geomorphological mapping and process monitoring.

The detrimental effects of wind erosion and dust-storms are recognized, their mechanisms are well appreciated, the human impact on the wind erosion system is understood, and effective control measures are available. The remaining wind erosion problem areas and events in the world today are therefore more a function of poor education and inadequate management than ignorance of the processes concerned.

REFERENCES

Anderson, R. (1987). Grassland revegetation by land imprinting a new option in desertification control, *Desertification Control Bulletin*, **14**, 38–44.

Armbrust, D. V. and Dickerson, J. D. (1971). Temporary wind erosion control: cost and effectiveness of 34 commercial materials, *J. Soil Wat. Conserv.*, **26**, 154–7.

Armbrust, D. V. and Lyles, L. (1975). Soil stabilizers to control wind erosion, *Soil Sci. Soc. Am. Spec. Publ.*, **7**, 77–82.

Bagnold, R. A. (1941). *The Physics of Blown Sand and Desert Dunes*, Methuen, London.

Bertrand, J., Cerf, A. and Domergue, J. L. (1979). Repartition in space and time of dust haze south of the Sahara, *WMO*, **538**, 409–15.

Bisal, F. and Hsieh, J. (1966). Influence of moisture on erodibility of soil by wind, *Soil Sci.*, **102**, 143–6.

Breed, C. S. and McCauley, J. F. (1986). Use of dust storm observations on satellite images to identify areas vulnerable to severe wind erosion, *Climatic Change*, **9**, 243–58.

Bryson, R. A. and Baerreis, D. A. (1967). Possibilities of major climatic modification and their implications: Northwest India, a case for study, *Bull. Am. Met. Soc.*, **48**, 136–42.

Burritt, B. and Hyers, A. D. (1981). Evaluation of Arizona's highway dust warning system, in *Desert Dust* (Ed. T. L. Péwé), pp. 281–92, Geological Society of America Special Paper 186.

Chakvetadze, E. A. (1962). Results of dust storm observations in the Irtysh River region, *Soviet Soil Sci.*, **4**, 180–6.

Changery, M. J. (1983). *A Dust Climatology of the United States*, NOAA.

Chepil, W. S. (1946). Dynamics of wind erosion: IV. The translocating and abrasive action of the wind. *Soil Sci.*, **61**, 167–77.

Chepil, W. S. (1951). Properties of soil which influence wind erosion, 4. State of dry aggregate structure, *Soil Sci.*, **72**, 387–401.

Chepil, W. S. (1955a). Factors that influence clod structure and erodibility of soil by wind, IV. Sand, silt and clay, *Soil Sci.*, **80**, 155–62.

Chepil, W. S. (1955b). Factors that influence clod structure and erodibility of soil by wind. V. Organic matter at various stages of decomposition, *Soil Sci.*, **80**, 413–21.

Chepil, W. S. and Milne, R. A. (1941). Wind erosion of soil in relation to roughness of surface, *Soil Sci.*, **52**, 417–33.

Chepil, W. S. and Woodruff, N. P. (1963). The physics of wind erosion and its control, *Adv. Agron.*, **15**, 211–302.

Choun, H. F. (1936). Dust storms in southwestern Plains area, *Mon. Weath. Rev.*, **64**, 195–9.

Claffin, L. E., Stuteville, D. L. and Armbrust, D. V. (1973). Windblown soil in the epidemiology of bacterial leaf spot of alfalfa and common blight of beans, *Phytopathology*, **63**, 1417–19.

Clements, T., Stone, R. O., Mann, J. F. and Eymann, J. L. (1963). A study of windborne sand and dust in desert areas, Technical Report ES-8, US Army Natik Laboratories, Earth Sciences Division, Natik, Mass.

Cooke, R. U. (1970). Stone pavements in deserts, *Annals Assoc. Am. Geographers*, **60**, 560–77.

Cooke, R. U., Brunsden, D. and Doornkamp, J. (1982). *Geomorphological Hazards in Urban Drylands*, Clarendon Press, Oxford.

Cooke, R. U. and Doornkamp, J. C. (1974). *Geomorphology in Environmental Management*, Clarendon Press, Oxford.

Das, T. M. (1988). Effects of deposition of dust particles on leaves of crop plants on screening of solar illumination and associated physiological processes, *Environmental Pollution*, **53**, 421–2.

Dickerson, J. D., Woodruff, N. P. and Banbury, E. E. (1976). Techniques for improving survival and growth of trees in semiarid areas, *J. Soil Wat. Conserv.*, **31**, 63–6.

Dresch, J. (1986). Degradation of natural ecosystems in the countries of Maghreb as a result of human impact, in *Arid Land Development and the Combat against Desertification* (Ed. M. H. Glantz), pp. 65–7, UNEP, Moscow.

Eckholm, E. P. (1976). Two costly lessons: the Dust Bowl and the Virgin Lands, in *Losing Ground*, pp. 46–57, Pergamon, Oxford.

FAO (1979). *A Provisional Methodology for Soil Degradation Assessment*, FAO, Rome.

FAO/UNEP/UNESCO (1980). *Provisional Map of Present Degradation Rate and Present State of Soil, Near and Middle East*, FAO, Rome.

Frank, A. B., Harris, D. G. and Willis, W. O. (1977). Growth and yields of spring wheat as influenced by shelter and soil water, *Agron. J.*, **69**, 903–6.

Fryberger, S. G. (1979). Dune forms and wind regime, in *A Study of Global Sand Seas* (Ed. E. D. McKee), pp. 136–69, US Geological Survey Professional Paper 1052.

Fryrear, D. W. (1981). Long-term effect of erosion and cropping on soil productivity, in *Desert Dust: Origins, Characteristics and Effects on Man* (Ed. T. L. Péwé), pp. 253–9, Geological Society of America Special Paper 186.

Gillette, D. A., Adams, J., Endo, A., Smith, D. and Kihl, R. (1980). Threshold velocities for input of soil particles into the air by desert soils, *Journal of Geophysical Research*, **85C**, 5621–30.

Goudie, A. S. (1983). Dust storms in space and time, *Progress in Physical Geography*, **7**, 502–30.

Grigoryev, A. A. and Kondratyev, K. J. (1980). Atmospheric dust observed from space, *WMO Bull.*, **29**, 250–5.

Gupta, J. P., Aggarwal, R. K. and Raikhy, N. P. (1981). Soil erosion by wind from bare sandy plains in western Rajasthan, India, *Journal of Arid Environments*,**4**, 15–20.

Gupta, J. P., Aggarwal, R. K., Gupta, G. N. and Kaul, P. (1983). Effect of continuous application of farmyard manure and urea on soil properties and production of pearl millet in western Rajasthan, *Indian J. Agric. Sci.*,**53**, 53–6.

Houseman, J. (1961). Dust haze at Bahrain, *Met. Mag.*, **90**, 50–2.

Jones, D. K. C., Cooke, R. U. and Warren, A. (1986). Geomorphological investigation, for engineering purposes, of blowing sand and dust hazard, *Quarterly Journal of Engineering Geology*, **19**, 251–70.

Kalma, J. D., Speight, J. G. and Wasson, R. J. (1988). Potential wind erosion in Australia: a continental perspective, *J. Climatol.*, **8**, 411–28.

Leathers, C. R. (1981). Plant components of desert dust in Arizona and their significance for man, in *Desert Dust* (Ed. T. L. Péwé), pp. 191–206, Geological Society of America Special Paper 186.

Lettau, K. and Lettau, H. H. (1978). Experimental and micrometeorological studies of dune migration, in *Exploring the World's Driest Climate* (Eds. H. H. Lettau and K. Lettau), pp. 110–47, IES Report 101, University of Wisconsin, Madison.

Lockeretz, W. (1978). The lessons of the Dust Bowl, *Am. Scient.*, **66**, 560–9.

Lyles, L., Tatarko, J. and Dickerson, J. D. (1983). Windbreak effects on soil water and wheat yield, American Society of Agricultural Engineers Paper 83, p. 2074.

McCalla, T. M. and Army, T. J. (1961). Stubble mulch farming, *Adv. Agron.*, **13**, 125–96.

McCauley, J. F., Breed, C. S., Grolier, M. J. and Mackinnon, D. J. (1981). The US dust storm of February 1977, in *Desert Dust* (Ed. T. L. Péwé), pp. 123–47, Geological Society of America Special Paper 186.

Middleton, N. J. (1985). Effect of drought on dust production in the Sahel, *Nature*, **316**, 431–4.

Middleton, N. J. (1986a). Dust storms in the Middle East, *Journal of Arid Environments*, **10**, 83–96.

Middleton, N. J. (1986b). A geography of dust storms in south-west Asia, *J. Climatol.*, **6**, 183–96.

Middleton, N. J. (1987). Desertification and wind erosion in the western Sahel: the example of Mauritania, School of Geography, Oxford, Research Paper 40.

Middleton, N. J. (1989). Desert dust, in *Arid Zone Geomorphology* (Ed. D. S. G. Thomas), Belhaven, London, pp. 262–83.

Middleton, N. J., Goudie, A. S. and Wells, G. L. (1986). The frequency and source areas of dust storms, in *Aeolian Geomorphology* (Ed. W. G. Nickling), pp. 237–59, Unwin Hyman, Boston.

Morgan, R. P. C. (1986). *Soil Erosion and Conservation*, Longman, Harlow.

Phillips, R. E., Blevins, R. L., Thomas, G. W., Frye, W. W. and Phillips, S. H. (1980). No-Tillage agriculture, *Science*, **208**, 1108–13.

Prospero, J. M. and Nees, R. T. (1977). Dust concentration in the atmosphere of the equatorial North Atlantic: possible relationship to the Sahelian drought, *Science*, **196**, 1196–8.

Prospero, J. M. and Nees, R. T. (1986). Impact of the North African drought and El Nino on mineral dust in the Barbados Trade winds, *Nature*, **320**, 735–8.

Pye, K. (1987). *Aeolian Dust and Dust Deposits*, Academic Press, London.

Rapp, A. (1974). *Review of Desertisation in Africa*, International Secretariat for Ecology, Stockholm.

Risser, J. (1985). Soil erosion problems in the USA, *Desertification Control Bulletin*, **12**, 20–5.

Schwab, G. O., Frevert, R. K., Edminster, T. W. and Barnes, K. K. (1966). *Soil and Water Conservation Engineering*, Wiley, Chichester.

Skidmore, E. L. (1986). Wind erosion control, *Climatic Change*, **9**, 209–18.

Skidmore, E. L. and Woodruff, N. P. (1968). Wind erosion forces in the United States and their use in predicting soil loss, Agricultural Handbook 346, Agricultural Research Service, US Department of Agriculture.

Smalley, I. J. (1970). Cohesion of soil particles and the intrinsic resistance of simple soil systems to wind erosion, *J. Soil Sci.*, **21**, 154–61.

Stallings, J. H. (1957). *Soil Conservation*, Prentice-Hall.

Troeh, F. R., Hobbs, J. A. and Donahue, R. L. (1980). *Soil and Water Conservation for Productivity and Environmental Protection*, Prentice-Hall.

UN (1977). *Desertification: Its Causes and Consequences*, Pergamon, Oxford.

UNEP (1985). *Desertification Control in Africa. Actions and Directory of Institutions*, Vol. 1, UNEP, Nairobi.

Wasson, R. J. and Nanninga, P. M. (1986). Estimating wind transport of sand on vegetated surfaces, *Earth Surf. Processes Landf.*, **11**, 505–14.

Wells, G. L. and Middleton, N. J. (1988). The alteration of land surface cover across the western Sahel recorded by orbital photography 1965–1986, paper presented at ISLSCP Second Results Meeting, Niamey, April, 1988.

Wicks, G. A. and Smith, D. E. (1973). Chemical fallow in a winterwheat–fallow rotation, *Weed Sci.*, **21**, 97–102.

Wilson, S. J. and Cooke, R. U. (1980). Wind erosion, in *Soil Erosion* (Eds. M. J. Kirkby and R. P. C. Morgan), pp. 217–51, Wiley, Chichester.

Woodruff, N. P. and Siddoway, F. H. (1965). A wind erosion equation, *Soil Sci. Soc. Am. Proc.*, **29**, 602–8.

Woodruff, N. P., Chepil, W. S. and Lynch, R. D. (1957). Emergency chiseling to control wind erosion, Kansas Agricultural Experimental Station Technical Bulletin 90.

Worster, D. (1979). *Dust Bowl*, Oxford University Press.

Yaalon, D. H. (1987). Saharan dust and desert loess: effect on surrounding desert soils, *Journal of African Earth Sciences*, **6**, 569–71.

CHAPTER 4

Soil Salinity—Causes and Controls

J. D. RHOADES

INTRODUCTION

Irrigation is an ancient practice of semi-arid and arid regions that predates recorded history. While only about 15% of the world's farmland is irrigated, it contributes about 35–40% of the supply of food and fibre, it stabilizes production against the vagaries of weather, and it permits agriculture in desert environments. With irrigation inevitably comes salination of soils and waters. The salt contained in the irrigation water tends to be left behind in the soil as the pure water passes back to the atmosphere through the processes of evaporation and plant transpiration. Typically, excess water is applied to the land with irrigation or enters it by seepage from delivery canals. These waters percolate through the soil and underlying strata (dissolving salts in the process) and flow to and cause waterlogging and salt-loading in lower-elevation lands. In turn, saline soils are formed there through the process of evaporation.

The salt problem in irrigated agriculture is not new. The rise and fall of the Mesopotamian civilization nearly six to seven thousand years ago has been attributed to the development of irrigated agriculture and subsequently to its failure as a result of rising water tables and soil salination. An estimated 30% of all irrigated land suffers from salt-caused yield reductions. Salinity threatens the economy of many arid countries, such as Egypt, Iraq, and Pakistan, where irrigation is the backbone of agriculture. Salinity also constitutes the most serious water quality problem in many arid and semi-arid rivers and groundwater systems. The problems of soil salination, waterlogging, and water pollution are increasing as irrigation is being expanded and less suitable waters and soils are being used to meet the ever-increasing need for food in the world.

Techniques for Desert Reclamation
Edited by A. S. Goudie
© 1990 John Wiley & Sons Ltd.

Surviving the salinity threat requires that the seriousness of the problem be widely recognized, that the processes contributing to salination be understood, and that effective control measures be developed and implemented which will sustain the viability of our irrigated agriculture. The effects of salts on plants and soils, the causes of salination, the extent of the problem, practices used to control salinity, and opportunities to beneficially use saline waters for irrigation are discussed in this chapter.

THE OCCURRENCE AND EXTENT OF SALINITY PROBLEMS

The major natural, saline regions of the world are found in poorly drained low-lying lands under semi-arid and arid conditions where large quantities of salts leached from higher regions have accumulated in the slowly flowing groundwater and basin sinks, where the water table is at or near to the soil surface, and where the salts have ascended into the soil due to the high evapotranspiration rate. A close relationship between the depth and salinity of the water table and the extent of salt accumulation in soils is established in natural, semi-arid regions for the reasons given above.

The impact of man on the circulation of salts has been profound. As a consequence of irrigation more water and salt have been applied to soils, more salt stored in the soil and deeper strata have been mobilized as more leaching and deep percolation have occurred, and the groundwater table has risen in many places. As a result, large areas of irrigated lands have become secondarily waterlogged and salinized, and associated surface waters have become increasingly salinized because of a reduction in their volume and through their reception of salt-laden drainage waters.

Nearly 10% of the total land area of the world is estimated to be sufficiently affected by salt to limit its utilization for crop production. These areas of salt-affected soils are widely distributed throughout the world; no continent is free from salt-affected soils (see Figure 4.1 and Table 4.1, after Szabolcs, 1985). Serious salt-related problems occur within the boundaries of at least seventy-five countries (Rhoades, 1988).

Thus, we see that salt-affected soils are found to occur under widely varying conditions of climate, geology, agriculture, and, of course, social and cultural systems. The economic and social repercussions of soil salination are felt most acutely by the populations of arid zones and mainly by less developed nations which depend primarily upon irrigated agriculture for their food production.

The increasing population of the world necessitates that the viability of the earth's soil and water resources be maintained in order to meet the increasing demand for food. The projected increase in croplands for the final quarter of the century is only 10%, yet the world demand for food is expected to approximately double according to the UN World Food Conference Report of 1974. In addition, agricultural lands are predicted increasingly to be

111

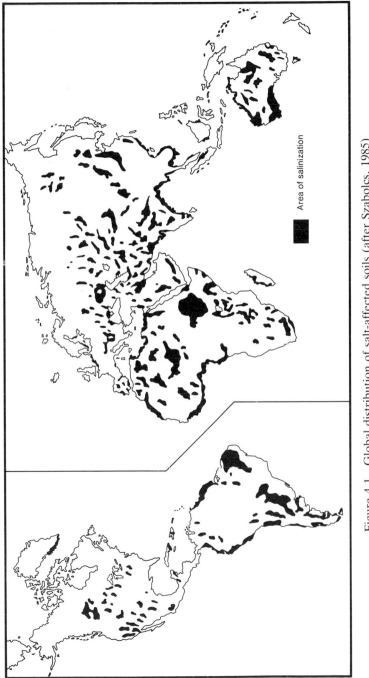

Figure 4.1 Global distribution of salt-affected soils (after Szabolcs, 1985)

Table 4.1   Extent of salt-affected soils by continents and
sub-continents

Region	Millions of hectares
Africa	80.5
Australasia	357.3
Europe	50.8
Mexico and Central America	2.0
North America	15.7
North and Central Asia	211.7
South America	129.2
South Asia	87.6
South East Asia	20.0
Total	954.8

diverted from agricultural production to other uses and increasingly to be
degraded through various means, a major one being salination. Associated
with the latter is the increasing pollution of water resources with various
chemicals and salts. Irrigated agriculture is heavily involved in these matters.

The amount of irrigated land increased from about 8 million hectares in the
year 1800 to 48 million hectares by 1900. It then approximately doubled in the
following 50 years and again doubled during the last 30 years. In some arid
countries, such as Egypt, nearly 100% of the agricultural land is irrigated,
while in others such as Pakistan it is about 50%. In less arid nations irrigated
land occupies a much lower proportion of the total cropland, but even there it
is continuing to increase and is reaching significant levels; for example, in
Thailand the percentage is 26, in France it is 13, in Spain 10, and in Greece
15. In the United States the area under irrigation doubled between 1949 and
1973 to 21 million hectares and by 1987 had more than doubled again. In the
Soviet Union about 1 million hectares of new irrigated land is developed each
year. In Hungary the irrigated area has increased tenfold since World War II.

The world's total irrigated area is estimated to be about 400 million
hectares by the year 2000. This increase will not only result in an increase in
world production in agriculture but also in an increase in water consumption,
in an increase in waterlogging of irrigated lands, and in an increase in salt
build-up in water supplies and irrigated lands. Unfortunately, no one has
predicted how much of this irrigated area will succumb to salt problems, but
based on past experience the problem of salt in irrigated lands will likely
increase at an even faster rate than that of the expansion of irrigation itself.

It is well known that large areas of the world (for example, old Mesopotamia, large parts of the Indus River Valley, vast territories in South
America and China, etc.) that previously supplied abundant crops by means

of irrigation have since succumbed to salination and waterlogging problems. For example, it is estimated that at one time Mesopotamia fed a population of between 17 and 25 million people and was a food exporter. Presently this area has a population only about one half what it once was and it imports a large quantity of food. People were forced to abandon the affected lands and to develop new areas. As long as new territories were available, the shifting of irrigated agriculture temporarily solved the problem. Today, however, with the growing density of population, increased degradation of land and water resources, and shrinkage of a suitable land base for agriculture, this practice of land abandonment is no longer generally acceptable.

In spite of the general awareness of these problems and past sad experiences, salination and waterlogging of our irrigated lands continues to increase. According to the estimates of the UN Food and Agriculture Organization and UNESCO, as much as half of the area of all existing irrigation systems of the world are seriously affected by salinity and/or waterlogging, the area potentially subject to secondary salination is estimated to be equal to or greater than the area presently affected, and 10 million hectares of irrigated land are abandoned yearly as a consequence of the adverse effects of salination and waterlogging. This phenomenon is common not only in old irrigation projects but also in areas where irrigation has only recently been introduced.

In some countries the salt problem threatens the national economy of that country. These problems are particularly serious in Argentina, Egypt, India, Iraq, Iran, Pakistan, and Syria. Roughly half the irrigated land in Syria's Euphrates River Valley has become saline to the point where crop losses now total an estimated $300 million annually. Between one-quarter and one-half of all irrigated land in South America is affected by salination, and the problem there is increasing. In India 35% of all irrigated land is seriously saline. In Pakistan, where 80% of all cropland is irrigated, one-third of it (about 6 million hectares) is experiencing severe salt problems and another 16% is threatened with salination by high water tables.

The future development of planned large irrigation projects, which involve diversions of rivers, construction of large reservoirs, and the irrigation of large land areas, has the potential to cause large changes in the water and salt balances and to affect the salinities of entire groundwater and river systems. The impact will certainly extend beyond that of the immediate irrigated area and can even affect neighbouring nations.

SALINITY EFFECTS ON PLANTS AND SOILS

Salt-affected soils and waters are those that are of reduced value for agriculture because of their content (or sometimes the past effects in the case of soils) of salts, consisting mainly of sodium, magnesium, calcium, chloride, and sulphate and secondarily of potassium, bicarbonate, carbonate, nitrate, and boron. Salts exert both general and specific effects on plants, thus influencing crop yields. Salts also affect soil physiochemical properties, which could in turn reduce the suitability of the soil as a medium for plant growth.

Effects on Plants

Excess salinity (essentially independent of its composition) in the root zone adversely affects plant growth by a general reduction in growth rate. The hypothesis that seems to best fit observations is that salt stress increases the energy that must be expended to extract water from the soil and to make the biochemical adjustments necessary to survive under stress. This energy is diverted from the processes that lead to normal growth and yield.

The salt tolerances of crops are expressed, after Maas and Hoffman (1977), in terms of their threshold values and percentage decrease in yield per unit increase of soil salinity in excess of the threshold (the preferred unit of soil salinity is the electrical conductivity of the extract of a saturated soil paste, EC_e, in dS/m). Salt tolerance data cannot indicate accurate, quantitative crop yield losses from salinity for every situation, since actual plant response to salinity varies with growing conditions, including climate, agronomic management, crop variety, etc. Salt tolerance data are useful, however, to predict how one crop might fare relative to another under similar conditions of salinity. Plants are generally most sensitive during the seedling stage; hence, it is imperative to keep salinity low in the seedbed. When salinity reduces plant stand, potential yields are decreased far more than that predicted by the salt tolerance data.

Typically, salt tolerance data apply most directly to surface-irrigated crops and conventional irrigation management. Sprinkler-irrigated crops may also suffer damage from foliar salt uptake and 'burn' from contact with the spray. Available data predicting yield losses from foliar spray effects are given in Maas (1986). The degree of foliar injury depends on weather conditions and water stress; for example, visible symptoms may appear suddenly when the weather becomes hot and dry.

Certain salt constituents are specifically toxic to some crops. Boron is highly toxic to many crops when present in the soil solution at concentrations of only a few parts per million (Maas, 1984b; Bingham *et al.*, 1985a,b). In some woody crops, sodium and chloride may accumulate in the tissue to toxic levels (Bernstein, 1974). The effects of salinity and toxic solutes on the physiology and biochemistry of plants are reviewed by Maas and Nieman (1978) and Maas (1984a).

Sodic soil conditions may induce calcium and various micronutrient deficiencies because of the associated high pH and bicarbonate conditions repressing their solubilities and concentrations. Sodic soils are less extensive than saline soils in irrigated lands.

Effects on Soils

The suitability of soils for cropping depends appreciably on the readiness with which they conduct water and air (permeability) and on aggregate properties that control the friability of the seedbed (tilth). In contrast to saline soils, sodic soils have lower permeabilities and poorer tilth, causing problems in many irrigated lands.

Because of negative electrical surface charges, clays absorb positively charged ions (cations), such as calcium, magnesium, and sodium, by electrostatic attraction. These cations can be replaced or exchanged by other cations that are added to the soil solution. Each soil has a measurable capacity to adsorb and exchange cations (the cation exchange capacity). The percentage of this capacity satisfied by sodium is referred to as the exchangeable sodium percentage (ESP). The percentage is approximately numerically equal to the sodium adsorption ratio (SAR) of the soil solution (SAR = $Na^+/[(Ca^{2+} + Mg^{2+})/2]^{1/2}$, where the concentrations are expressed in nmol(+)/litre). Thus, SAR can be used essentially interchangeably with ESP over the normal range of ESP encountered in irrigated soils.

The adsorbed ions in the 'envelope' around colloidal clay are subject to two opposing processes: (a) they are attracted to the negatively charged clay surface by electrostatic forces and (b) they tend to diffuse away from the surface of the clay under a concentration gradient. The two opposing processes result in an approximately exponential decrease in adsorbed ion concentration with distance from the clay surface out into the bulk solution. Divalent cations, such as calcium and magnesium, are attracted to the surface with a force twice as great as monovalent cations, like sodium, for example. Thus, the 'envelope' in the divalent system is more compressed towards the clay surface. The 'envelope' is also compressed by an increase in the electrolyte concentration of the bulk solution.

Short-range adhesive forces, called van der Waals forces, are involved in the particle-to-particle associations that bind the clays into aggregates. The net forces, which result in the formation of aggregates, are diminished when the cation 'envelopes' are extended and are enhanced when they are compressed. This occurs because the relatively long-range electrostatic charged 'envelopes' around adjacent clay particles repel one another. With compression of the cation 'envelope' towards the clay surface, the overlap of the 'envelopes' of two adjacent particles is reduced for a given distance between them, the repulsion forces between the like-charged 'envelopes' are decreased, and the particles can approach sufficiently close to permit the van der Waals forces to come into play. The resulting aggregate structure is more porous, resulting in enhanced permeability and tilth. When repulsion between clay particles is predominant, more solution is imbibed between clay particles, causing swelling. Such swelling reduces the size of the interaggregate pore spaces in the soil and hence reduces permeability. Swelling is primarily important in soils that contain expanding layer phyllosilicate minerals (smectites like montmorillonite) and ESP values in excess of about 15. For such minerals, exchangeable sodium is initially preferentially adsorbed on the external clay surfaces. These external surfaces make up about 15% of the cation exchange capacity (CEC). Only with further 'build-up' of adsorbed sodium does it enter the interlayer position between the parallel platelets of the oriented and associated clay particles of the sub-aggregate assemblages, called domains, where it creates the repulsion forces that lead to swelling. Dispersion (release of individual clay platelets from aggregates) and slaking (breakdown of aggregates into sub-aggregate assemblages) can occur at ESP values lower than 15, providing the electrolyte concentration is sufficiently low. Dispersed platelets or slaked sub-aggregate units can lodge in pore interstices, also reducing permeability. Soil solutions composed of high solute concentrations and calcium and magnesium salts produce good soil physical properties. Conversely, low concentrations and sodium salts adversely affect permeability and tilth.

When water infiltrates the soil surface, the soil solution of the topsoil is essentially that of the infiltrating water, while the exchangeable sodium percentage is essentially that preexistent in the soil (since ESP is buffered against rapid change by the soil CEC). All water entering the soil must pass through the surface; hence, the stability of the topsoil aggregates influences the water entry rate of the soil. Representative guideline threshold values of SAR (~ESP) and the electrical conductivity of infiltrating water for maintenance of soil permeability are given in Rhoades (1982). Effects of salts on soil properties are reviewed by Keren and Shainberg (1984), Shainberg (1984), Shainberg and Letey (1983), and Emerson (1984).

SOURCES OF SALT AND CAUSES OF SOIL SALINATION

The original sources of salts are the dissolved products of mineral weathering, emanations from volcanic eruptions, discharges from deep thermal sources, and the primary ocean. These salts have been redistributed over time. Winds blowing over the oceans pick up salt particles, which originate at the sea surface as spray, and carry many of them onto the land where they are mixed with other salts derived from weathering products and sedimentary sources.

As a result of certain conditions of climate, topography, geologic history, land use, or the nature of the sediment or soil, salts have accumulated in certain locations in amounts many times higher than the average concentration. In landscapes with good rainfall and effective drainage systems, soluble salts are transported by flowing surface and ground waters eventually to the sea. During this migration their concentration and composition undergo many changes as a result of their different mobilities and affinities to form or interact with compounds they meet in their path. The migration and redistribution is essentially exclusively through the agency of water, which acts both as solvent and transporting vehicle. However, in many parts of the world with internal or ineffective drainage the salts accumulate in relatively low-lying regions such as valley basins or upland depressions. Such obvious areas of accumulation account for only a fraction of the salts in the landscape. Much salt is stored in sub-soils and deeper sub-strata of the hydrogeologic system as well as in groundwaters. In some regions marine incursions in the past have left buried saline sediments in the landscape; often these underlie irrigation projects or rain-fed agricultural lands. Such salt reservoirs may be returned to circulation after a change in the local topographic or climatic conditions or through the actions of man.

Salt-affected soils occur mostly in regions of an arid or semi-arid climate, that is where evapotranspiration exceeds rainfall and, hence, where leaching and transportation of salts to the oceans is not so complete as in humid regions. Such soils also usually occur in relatively low-lying places that receive water by gravitational flow from higher locations. Sodic soils usually occur in slightly elevated areas that receive salt inputs from the upward, capillary flow of soil water and are often found adjacent to saline and periodically waterlogged areas; sodium accumulates there because of its comparatively high solubility and mobility. These slightly elevated areas are periodically leached by rain or snowfall which, at least temporarily, reduces the concentration of soluble salts in them.

Restricted drainage usually contributes to the salination of soils and may involve low permeability of the soil or the presence of a high groundwater table. High groundwater tables are often related to the topographic position. The drainage of waters from the higher lands of valleys and basins may raise the groundwater level so that it is near the soil surface in the lower lands. Low permeability of the soil causes poor drainage by impeding the downward movement of water.

While salt-affected soils occur extensively under natural conditions, the salt problems of greatest importance in agriculture arise when previously productive soil becomes salinized as a result of irrigation or removal of natural vegetation and certain dryland agricultural practices (so-called secondary salination). Man's activities have increased salt-affected areas considerably either by adding more water by irrigation or by using less, as when dryland agriculture replaces native vegetation. In either case water infiltrated into the soil in excess of that used by the agricultural crops passes beyond the root zone, picking up salts from the soils and sub-strata and often developing waterlogged areas in low areas where those waters flow. When this occurs, soluble salts stored in the ground are mobilized to accumulate at the surface in the seepage areas, salinizing the soils where the rising water tables approach ground level and increasing solute concentrations in associated groundwaters and streams. Irrigated agricultures' role in salinizing soils and water systems has been well recognized for hundreds of years. However, it is only relatively recently recognized that the clearing of lands for dryland agriculture has created analogous problems. The latter problem occurs even in areas such as Australia, where the level of soil salinity under natural conditions is typically very low. Also it is of relatively recent recognition that salination of water resources from agricultural activities is a major and widespread phenomenon of likely even greater concern than that of the salination of soils. Only in the past few years has it become apparent that trace toxic constituents, such as selenium, in agricultural drainage waters can cause serious pollutional problems.

RECLAMATION OF SALT-AFFECTED SOILS

There are three principal aspects of the salt problem and its control in irrigated agriculture. One is the improvement (reclamation) of soils that are salt-affected under natural conditions or have become so because of mis-management. A second aspect is the management of productive or only slightly salt-affected soils so as to prevent an increase in their salinity and reduction in crop yields. A third aspect is management to minimize the pollution of ground and surface water supplies with salts and chemicals as a consequence of irrigated agriculture.

Saline soils are reclaimed by improving drainage and by leaching with irrigation water to remove excess salts. The improvement of sodic soils involves (besides drainage and leaching) the replacement of excessive adsorbed sodium by calcium or magnesium and of practices that develop better soil structure and permeability. Guidelines and procedures for reclaiming salt-affected soils are described in detail in US Salinity Laboratory Staff (1954), Rhoades (1982), and Rhoades and Loveday (1990).

Adequate drainage is essential for the permanent improvement of salt-affected soils. In order to prevent waterlogging, drainage must remove the precipitation and irrigation water infiltrated into the soil that is in excess of crop demand and any other water that seeps into the area; it must provide an outlet for the removal of salts that accumulate in the rootzone in order to avoid soil salination; and it must keep the water table sufficiently deep to prevent the flow of salt-laden groundwater up to the rootzone by capillary forces. Drainage systems are essentially engineering structures that remove water according to the principles of soil physics and hydraulics. New materials, new methods of installation, and the use of larger and more powerful machinery have revolutionized this industry in recent years, so that drainage facilities can now be constructed much more easily, quickly, and precisely than ever before. Typically, plastic drain tubes enveloped by synthetic 'filtersocks' are 'plowed-in' at the desired depth and grade. This 'plow' is precisely and automatically controlled by a laser-guidance system that is an integral part of a relatively fast-moving, self-propelled drain-installation tractor unit. Computer models that can simulate water table levels and salt removal under alternative conditions of cropping and water management are available to better assess and design the drainage needs of the area. Various tillage equipment can even invert whole soil profiles or break up sub-strata as deep as 2.5 m that impede deep percolation, so that many adverse physical soil conditions causing or associated with salt-affected soils can be modified so as to improve their leachability and drainability. For more information on drainage for salinity control see van Schilfgaarde (1974, 1984) and Rhoades (1974).

Once drainage is provided, saline soils are reclaimed by applying water to the soil surface and allowing it to pass downward through the rootzone. Leaching efficiency has been greatly increased through improvements in the accuracy and precision of land-levelling techniques and by the ability to apply water uniformly across the field. New theories and guidelines have been developed to predict the amounts of water needed to reduce the soluble salts for various conditions of soil properties and methods of water application (Rhoades, 1982). A better understanding of the chemistry of soil permeability has been developed and quicker and more cost-effective procedures have been developed for reclaiming sodic soils (Rhoades and Loveday, 1990).

SALINITY-RELATED PROCESSES OPERATIVE IN SOIL–PLANT–WATER SYSTEMS

Salinity management requires an understanding of not only how salts affect plants and soils but also of how cropping and irrigation affect soil and water salinity.

Irrigation–Evapotranspiration–Leaching–Drainage Interactions

The concentrations of soluble salts increases in soils as the applied water, but not salts, is removed by evaporation and transpiration. Evapotranspiration (ET) can cause an appreciable upward flow of water and salt into the rootzone from lower soil depths. By this process, many soils with shallow, saline water tables become salinized. Soluble salts will eventually accumulate in irrigated soils to the point where crop yields will suffer unless steps are taken to prevent it. To prevent the excessive accumulation of salt in the rootzone, irrigation water (or rainfall) must be applied in excess of that needed for ET and must pass through the rootzone to leach out the accumulating salts. This is referred to as the 'leaching requirement' (L, the fraction of infiltrated water that passes through the rootzone; US Salinity Laboratory Staff, 1954). Once the soil solution has reached a salinity level compatible with the cropping system, subsequent irrigations must remove at least as much salt from the rootzone as is brought in with irrigations, a process called 'maintaining salt balance'. In fields irrigated to steady-state conditions with conventional irrigation management, the salt concentration of the soil water is essentially uniform near the soil surface regardless of the leaching fraction, but increases with depth as L decreases. Likewise, average rootzone salinity increases and crop yield decreases as L decreases. Details on methods to calculate the leaching requirement and salt balance are given by Rhoades (1974, 1982).

Adequate drainage is mandatory to handle the leachate needed to achieve the leaching requirement and salt balance. In addition, the water table depth must be controlled to prevent any appreciable upward flow of water and salt into the rootzone. This water table depth is irrigation management dependent and not single valued as is commonly assumed (van Schilfgaarde, 1976).

Soil Salinity–Plant Interactions

The time-average rootzone salinity is affected by the degree to which the soil water is depleted between irrigations (Rhoades, 1972). As the time between irrigations is increased, the matric potential decreases as the soil dries, and the osmotic potential decreases as salts concentrate in the reduced water volume. Crop yield is closely related to the time- and depth-averaged total soil water potential, i.e. matric plus osmotic (Ingvalson *et al.*, 1976). As water is removed from a soil with non-uniform salinity distribution, the total water potential of the water being absorbed by the plant tends to approach uniformity in all depths of the rootzone (Rhoades and Merrill, 1976). Following irrigation, plant roots absorb water in soil depths of low osmotic stress rather than in regions of high osmotic stress. Normally this means that most of the water uptake is from the upper, less saline soil depths until sufficient water is removed to equalize the total water stress with depth. After that, salinity effects on crop growth will be magnified. This implies that: (a) plants can tolerate higher levels of salinity under conditions of low matric stress (e.g. high-frequency forms of irrigation, like drip) and (b) high soil-water salinities occurring in deeper regions of the rootzone can be significantly offset if sufficient low-salinity water is added to the upper profile fast enough to satisfy the crop's evapotranspiration requirement. Research results tend to support these conclusions (van Schilfgaarde *et al.*, 1974). Thus, irrigation management affects permissible levels of salinity of soils and irrigation waters. A typical deficiency of prevalent classification schemes of water quality for irrigation is that they exclude irrigation management effects. For more appropriate methods of assessing water suitability for irrigation, see Rhoades (1972, 1982, 1984a, 1984b) and Rhoades and Merrill (1976).

Soil Salinity–Irrigation System Interactions

The distribution within and the degree to which a soil profile becomes salinized are also functions of the degree and manner of water application and leaching. More salt is generally removed per unit of leachate with sprinkler irrigation than with flood irrigation. Thus, the salinity of water applied by sprinkler irrigation could be higher than that applied by flood or furrow irrigation with a comparable degree of cropping success, provided foliar burn is avoided. There is evidence that trickle irrigation, in which water is applied steadily at a rate slightly in excess of ET, permits crops to be grown more successfully with saline waters than otherwise possible. With this method, the high matric potential resulting from the high soil water content and limited drying between irrigations minimizes time-averaged soil-water salinity. Crop salinity tolerances determined under flood and furrow irrigation may not be directly applicable to trickle irrigation because of the higher water potential achieved with the latter form of irrigation; however, substantive data are lacking in this regard.

As noted above, the salt-removal efficiency with sprinkler irrigation tends to be substantially higher than with flood and trickle irrigation. Solute transport is governed by the combined processes of convection (movement with the bulk solution) and diffusion (movement under a concentration gradient); convection is usually the predominant process.

Differential velocities of water flow normally occur within the soil matrix (dispersion) because the pore size distribution is typically non-uniform. Dispersion is appreciable when flow velocity is high, and diffusion often limits salt removal under such conditions. Soils with large cracks and well-developed structure are especially variable in their water and solute transport properties because the large 'pores' are preferred pathways, as are earthworm channels, old root holes, interpedal voids, etc., and most of the flow in flooded soils occurs in them. Much of the water and salt in intra-aggregate pores is 'bypassed' in flood-irrigated soils. Flow velocity and water content are typically lower in soils irrigated with sprinklers; hence, bypass is reduced and efficiency of salt leaching is increased. For a more quantitative description of effects of convection and dispersion on solute transport in soils see the review of Wagenet (1984).

Salinity–Soil Interactions

Other soil-related processes also affect salt concentration and transport during the irrigation and leaching of soils. In most arid land soils, the clay particles are dominated by negative charges, which can retard cation transport through exchange processes. Simultaneously, anions are effectively excluded from part of the pore solution adjacent to the negatively charged clay surface, accelerating their transport. Boron also undergoes absorption reactions that retard its movement. These reactions are reviewed by Wagenet (1984).

Dissolution and precipitation of salts and mineral weathering significantly affect the composition of the soil solution and salt-loading contributions from irrigation. Studies by Rhoades *et al.* (1973, 1974) have shown that the effects of salt precipitation are generally insignificant at leaching fractions of 0.2 and higher with irrigation waters of less than about 1.0 dS m electrical conductivity (EC_{iw}). At leaching fractions of 0.1 or less, salt precipitation is frequently significant, depending on the composition of the irrigation water. For waters with an EC_{iw} of less than ~0.4 dS m, the dissolution of minerals is often more important in controlling the levels of soil-water salinity than is the salt content of the irrigation water. Models of soil chemistry have been developed and coupled to descriptions of solute transport (Wagenet, 1984). These models are primarily based on chemical equilibrium concepts and seldom include silicate mineral weathering. Furthermore, many salt dissolution/precipitation reactions are kinetically controlled in soil systems. Lack of appropriate descriptions of mineral weathering and other kinetically controlled reactions in irrigated soils are major factors limiting the validity of prevalent chemistry models (Jurinak, 1984).

The hydraulic properties of the soil depend upon total salt concentration of the percolating water and the nature of the adsorbed cations. The sodium adsorption ratio of the soil water is a good estimate of the exchangeable sodium percentage of the soil, as discussed earlier, and is frequently used for diagnosing sodicity problems. However, SAR_{sw} is related to but is not the same as that of the irrigation water SAR_{iw}. Changes in SAR_{iw} occur as the irrigation water infiltrates the soil because of concentration by ET, the accumulation of salts in the seedbed, and because of evaporation and the decomposition of plant residues near the surface. Other factors affecting SAR_{sw} are the loss or gain in Ca and Mg salts due to precipitation of alkaline earth carbonates present in the irrigation water and the introduction of Ca, Mg, and HCO_3 into the soil water from the dissolution and weathering of soil minerals. These effects limit the applicability of SAR_{iw} as a suitable index of SAR_{sw} to relatively saline, low carbonate waters. For sodic waters, the more generally applicable adjusted sodium adsorption ratio should be used in its place. The adjusted SAR_{sw} may be calculated by either of two methods, which give essentially equivalent results, as described elsewhere (Rhoades, 1982, 1984b; Suarez, 1981, 1982; Oster and Rhoades, 1977). Soil permeability will be reduced if the adjusted SAR_{sw}–EC_{iw} combination lies to the left of a threshold relation between adjusted SAR_{sw} (ordinate) and EC_{iw} (abscissa). The threshold relation curves downward below adjusted SAR_{sw} values of 10 and intersects the EC_{iw} axis at a value of about 0.3 because of the dominating effect of electrolyte concentration on soil aggregate stability, dispersion, and crusting at such low salinities. A representative SAR_{sw}–EC_{iw} threshold relation is given in Rhoades (1982) as a guideline for arid-land soils.

In many semi-arid regions the irrigation season is followed by a rainy season. During the irrigation season the salinity of the irrigation water usually prevents excessive aggregate slaking, soil swelling, and clay dispersion. When the irrigation water is displaced by rainwater, a SAR_{sw}–EC_{iw} situation conducive to disaggregation, dispersion, and crusting can result. Insufficient research has been directed towards prediction of this type of response, with resulting limitations in the management of salt-affected soils. Indeed, it is a function not only of soil sodicity but also of other soil properties, including the rates of soil mineral weathering, salt dissolution and transport, and cation exchange. For more information on this topic, see Shainberg (1984).

Adsorption by the soil of some solutes like boron also occurs with the irrigation process. Plants response primarily to the boron concentration of the soil water rather than to the amount of adsorbed B (Keren *et al.*, 1985a, 1985b; Bingham *et al.*, 1985a, 1985b). Some boron added with the irrigation water will be adsorbed by the soil, but boron still concentrates in the soil water. For some transitional period of time, the degree to which boron is concentrated in the soil water will be less than that of non-adsorbed solutes like chloride. The time required to reach a state when boron concentration in the soil water reaches its maximum is typically three to five years, but varies with soil properties, amount of irrigation water applied, leaching fraction, and concentration of B in the irrigation waters.

The prevalent models of solute reactions and transport in irrigated soils suffer the deficiency of not appropriately representing the large variations in the above-described processes that often occur under field conditions. Only recently has this problem been approached directly by measuring, on a large scale, solute distributions in field soil profiles. The results to date indicate that we do not yet have a suitable method to summarize and to integrate the processes operative on a field basis (Jury, 1984). It is probable that alternative modelling approaches, like that proposed by Corwin *et al.* (1988, 1989), may help in this regard.

SALINITY-RELATED PROCESSES OPERATIVE IN IRRIGATION PROJECTS AND GEOHYDROLOGIC SYSTEMS

Some unique effects of irrigation are operative at the scale of whole projects and entire geohydrologic systems; hence, some management practices for salinity control should address this larger scale.

Irrigation Return Flow

The primary sources of irrigation return flow are bypass water, canal seepage, deep percolation, and tailwater or surface runoff. Bypass water is often required to maintain hydraulic head and adequate flow through the canal system. It is usually returned directly to the river, and few pollutants are picked up by this route. Canal seepage may contribute to high water tables, increase groundwater salinity and phreatophyte growth, and generally increase saline drainage from irrigated areas. Law *et al.* (1972) estimated that 20% of the total water diverted for irrigation in the United States is lost by seepage from conveyance and irrigation canals. If the water passes through salt-laden sub-strata or displaces saline groundwater, the salt pickup from this source can be substantial. An example is the Grand Valley of Colorado. Canal lining can reduce such salt loading. Evaporation losses from canals commonly amount to only a small percentage of the diverted water. Closed conduit conveyance systems can minimize both seepage and evaporation losses and ET by phreatophytes. The closed conduit system also provides the potential for higher project irrigation efficiency and lower salt loading (van Schilfgaarde and Rawlins, 1980).

Salt Loading from Irrigation and Drainage

Irrigation water may contain from 0.05 to 3.5 tons of salt per $1000\,m^3$. With crops requiring annual irrigations of 6000 to $9500\,m^3$ water per hectare to meet ET, from 0.3 to 32 tons/ha of salt may be added to irrigated soils annually. Reducing the volume of water applied will reduce the amount of salt added and the amount to be removed by leaching. Minimizing the leaching fraction maximizes the precipitation of applied Ca, HCO_3, and SO_4 salts as carbonates and gypsum minerals in the soil, and it minimizes the 'pickup' of weathered and dissolved salts from the soil. The salt load from the rootzone can be reduced from about 2 to 12 tons/ha per year by reducing L from 0.3 to 0.1 (Rhoades *et al.*, 1973, 1974; Rhoades and Suarez, 1977; Oster and Rhoades, 1975).

Minimizing leaching may or may not reduce salinity degradation where the drainage water is not intercepted and is returned to the associated surface water or groundwater. A reduction of degradation will generally occur where saline groundwaters with concentrations in excess of those of the recharging drainage waters are displaced into the surface water. Many such situations occur in the upper Colorado River basin. Reduced leaching will not reduce salinity of the receiving water if it is already saturated with these constituents. Rivers unsaturated with gypsum but essentially saturated with $CaCO_3$ will not benefit from reduced leaching unless salts other than those derived from the diverted water or from soil mineral weathering and dissolution in the rootzone are encountered in the drainage flow path or a 'foreign' saline groundwater is displaced by the drainage water to the river. The Colorado River in its lower basin is probably of this type.

Like surface waters, groundwater receiving irrigation drainage water may not benefit from reduced leaching. With no sources of recharge other than drainage return flow, the groundwater eventually must come to the composition of the drainage water, which will be more saline with low leaching. However, the groundwater salinity may be lower with reduced leaching for an interim period of time. For groundwater being pumped for irrigation with no recharge other than by drainage return, the short-term limitations are the same as described above. Groundwater undersaturated with $CaCO_3$ (unlikely in arid lands) will show a slight benefit under low leaching, groundwater saturated with $CaCO_3$ will show no benefit under low leaching, and groundwater saturated with $CaCO_3$ and nearing saturation with gypsum will show substantial benefit from low leaching. Low leaching management can continuously reduce degradation of the groundwater, only if other sources of high-quality recharge into the basin exist and if flow out of the basin is high relative to drainage inflow. If a fixed volume of saline water is disposed of in a closed basin by irrigation, groundwater salinity will usually be lower with high leaching (Rhoades and Suarez, 1977).

The extent to which leaching can be minimized is limited by the salt tolerances of the crops being grown. In most irrigation projects, the currently used L's can be reduced appreciably without harming crops or soils, especially with improvements in irrigation management (van Schilfgaarde *et al.*, 1974).

PRACTICES TO CONTROL SALINITY IN THE ROOTZONE

Management practices for the control of salinity and sodicity include: selection of crops or crop varieties that will produce satisfactory yields under the existing conditions of salinity or sodicity; use of land-preparation and tillage methods that aid in the control or removal of salinity; special planting procedures that minimize salt accumulation around the seed; irrigation to maintain a relatively high level of soil moisture and to achieve periodic leaching of the soil; and special treatments (such as additions of chemical amendments, organic matter, and growing green manure crops) to maintain soil permeability and tilth. The crop grown, the quality of water used for irrigation, and soil properties determine to a large degree the kind and extent of management practices needed.

Growing Suitably Tolerant Crops

Where salinity cannot be entirely eliminated, the judicious selection of crops that can produce satisfactory yields under moderately saline conditions is required. In selecting crops for saline soils, particular attention should be given to the salt tolerance of the crop during seedling development, because poor yields frequently result from failure to obtain a satisfactory stand. Some crops that are salt tolerant during later stages of growth are quite sensitive to salinity during early growth. Among the highly tolerant crops are barley, sugar beet, cotton, bermuda grass, Rhodes grass, western wheatgrass, bird's-foot trefoil, table beet, kale, asparagus, spinach, and tomatoes. Crops having low salt tolerance include radishes, celery, beans, clovers, and nearly all fruit trees (Maas and Hoffman, 1977).

Managing Seedbeds and Fields to Minimize Local Salinity Accumulation

Failure to obtain a satisfactory stand of furrow-irrigated row crops on moderately saline soils is a serious problem in many places. The failures are usually due to the accumulation of soluble salt in raised beds that are 'wet-up' by furrow irrigation. Modifications in irrigation practices and bed shape may reduce salt accumulation near the seed. The tendency of salts to accumulate near the seed during irrigation is greatest in single-row, round-topped beds. Sufficient salt to prevent germination may concentrate in the seedzone, even if the average salt content of the soil is moderately low. With double-row, flat-topped beds, however, most of the salt moves into the centre of the bed, which leaves the shoulders relatively free of salt, thus enhancing seedling establishment. Sloping beds are best for saline soils because the seed can be safely planted on the slope below the zone of salt accumulation. Planting in furrows or basins is satisfactory from the standpoint of salinity control but is often unfavourable for the emergence of many row crops because of crusting or poor aeration. Preemergence irrigation by sprinklers or by special furrows placed close to the seed may be used to keep the soluble salt concentration low in the seedbed during germination and seedling establishment. After the seedlings are established, the special furrows may be abandoned and new furrows made between the rows, or sprinkling may be replaced by furrow irrigation.

Careful levelling of land makes possible a more uniform application of water and, hence, better salinity control. Barren or poor areas in otherwise productive fields often are high spots that do not receive enough water for good crop growth or for leaching purposes. Lands that have been irrigated one or two years after initial levelling often need to be replaned to remove the surface unevenness caused by the settling of fill material. Annual crops should be grown after the first levelling so that replaning can be performed before a perennial crop is planted.

Irrigating to Maintain High Soil Water Potential and to Periodically Leach Salts

The method and frequency of irrigation and the amount of irrigation water applied may be managed to control salinity. The main ways to apply water are basin flooding, furrow irrigation, sprinkling, sub-irrigation, and drip irrigation. Flooding the entire surface is suitable for salinity control if the land is level, though aeration and crusting problems may occur. Aeration and crusting problems are minimized with furrow irrigation, but salts tend to accumulate in the beds. If excess salt does accumulate, a rotation of crops and periodic irrigation by flooding is a possible salinity-control measure. Alternatively, cultivation and irrigation depths can be modified, once the seedlings are well established, to 'shallow' the furrows so that the beds will be leached by later irrigations. Irrigation by sprinkling may permit better control of the amount and distribution of water. The tendency is to apply too little water by this method, and leaching of salts beyond the rootzone is accomplished only with special effort. Salinity is kept low in the seedbed during germination, but crusting may be a problem. Sub-irrigation, in which the water table is maintained close to the soil surface, is not generally suitable when salinity is a problem unless the water table is lowered periodically and leaching of the accumulated salts is accomplished by rainfall or by surface applications of water. Drip irrigation, if properly designed, minimizes salinity and matric stresses because the soil water is kept relatively high and salts are leached to the periphery of the wetted area. As noted earlier, higher levels of salinity in the irrigation water can be tolerated with drip as compared with other methods of irrigation.

Because soluble salts reduce the availability of water in almost direct proportion to their total concentration in the soil solution, irrigation frequency, irrespective of method of irrigation, should be increased so that the moisture content of saline soils is maintained as high as is practicable, especially during seedling establishment and the early stages of vegetative growth.

Managing Soils to Sustain Tilth

Sodic soils are especially subject to puddling and crusting. They should be tilled carefully, taking care to avoid wet soil conditions. Heavy machinery traffic should also be avoided. More frequent irrigation, especially during the germination and seedling stages of plants, tends to soften surface crusts on sodic soils and encourages better stands. Amendments such as gypsum, organic matter, and animal and green manures may be used to maintain permeability and tilth.

PRACTICES TO CONTROL SALINITY IN IRRIGATED LANDS

Improvements in the efficiencies of the delivery and application systems will appreciably facilitate salinity control in irrigated lands. Overirrigation contributes to the water table and salinity problems and increases the amount of water that the drainage system must accommodate. Therefore, a proper relation between irrigation, leaching, and drainage must be maintained in order to prevent irrigated lands from becoming salt affected. The amount of water applied should be sufficient to supply the crop and satisfy the leaching requirement but not enough to overload the drainage system. Overirrigation is a major cause of salinity build-up in many irrigation projects of the world.

Operating Delivery Systems Efficiently

Excessive loss of irrigation water from canals constructed in permeable soil is a major cause of high water tables and saline soils in many irrigation projects. Such seepage losses should be reduced by lining the canals with impermeable materials or by compacting the soil to achieve a very low permeability. Because the amount of water passing critical points in the irrigation delivery system must be known in order to provide water control and to achieve high water-use efficiency, provisions for effective flow measurement should be made. Unfortunately, many current irrigation systems do not use flow measuring devices and, thus, the individual farmers operate their own turnout facilities with limited control of the amount diverted to the farms. In addition, many delivery systems encourage overirrigation because water is supplied for fixed periods, or in fixed amounts, irrespective of seasonal variations in on-farm needs. Salinity and water table problems are often the result. Increasing the efficiency of the distribution system to provide water on demand and in metered amounts facilitates salinity control.

Irrigating Efficiently

Improvements in salinity control come with improvements in on-farm irrigation efficiency. The key to effective irrigation and salinity control is to provide the proper amount of water at the proper time. The optimum irrigation scheme provides water continuously to keep the soil-water content in the rootzone within narrow limits, although carefully programmed periods of stress may be desirable to obtain maximum economic yield with some crops; cultural practices also may demand periods of dry soil. Thus, careful control of timing and amount of water applied is a prerequisite to good water use efficiency and to high crop yield, especially in saline soils. As mentioned above, this requires water delivery to the field on demand which, in turn, requires close coordination between the farmer and the entity that distributes the water; it calls for devices to measure water flow (rates and volumes), and feedback devices that measure the water and salt content in the soil.

Automated solid-set and centre-pivot sprinkler systems are conducive to good control and distribution; in principle, trickle irrigation is even better. Gravity systems, if designed and operated properly, can also achieve good control. Laser-controlled precision land levelling allows better areal water distribution over the field and smaller water applications; combined with automation, it has led to high irrigation efficiencies for dead-level, flooded systems (Dedrick *et al.*, 1978). Using closed conduits, rather than open waterways for laterals, has the advantage of effective off/on control, in addition to capturing gravitational energy for pressurizing delivery systems or controls. In furrow-irrigated areas, furrow length can be reduced—and thus intake distribution is improved and tailwater eliminated—using a system such as Worstell's (1979) multiset system. Surge irrigation can improve irrigation uniformity in graded furrows (Bishop *et al.*, 1981). For tree crops, a low-head bubbler system that provides excellent control while minimizing pressure requirements has been developed (Rawlins, 1977). Drip systems, of course, are increasingly being used for permanent crops and high-value annual crops. Numerous opportunities exist for modifying existing irrigation systems to increase the effectiveness of water and salinity control.

A frequent constraint in improving on-farm water use is the lack of knowledge of just when irrigation is needed and of how much capacity for replenishment is available in the rootzone. Ways to detect the onset of plant stress and to determine the amount of depleted soil water are prerequisites to supplying water on demand and in the amount needed. Prevalent methods of scheduling irrigation usually do not, but should incorporate salinity effects on water availability (Rhoades *et al.*, 1981). Irrigation management for salinity control is the subject of reviews by van Schilfgaarde (1976) and van Schilfgaarde and Rawlins (1980).

PRACTICES TO CONTROL SALINITY IN WATER RESOURCES

Irrigated agriculture is a major contributor to the salinity of many rivers and groundwaters. The agricultural community has a responsibility to protect the quality of these waters. It must also maintain a viable, permanent irrigated agriculture. Irrigated agriculture cannot be sustained without adequate leaching and drainage to prevent excessive salination of the soil, yet these processes are the very ones that contribute to the salt loading of our rivers and groundwaters. River and groundwater salinity could be reduced if salt loading were minimized or eliminated. The protection of our water resources against excessive salination, while sustaining agricultural production through irrigation, will require a comprehensive land- and water-use policy that incorporates the natural processes involved in the soil–plant–water and associated geohydrological systems.

Strategies to consider in coping with increasing salinity in receiving water systems resulting from irrigation include: (a) eliminating irrigation, (b) intercepting point sources of drainage return flow and diverting them to other uses, and (c) reducing the amount of water lost in seepage and deep percolation.

Minimizing Deep Percolation and Intercepting Drainage

Deeply percolating water often displaces saline groundwater of higher salinity or dissolves additional salt from the sub-soil. Reducing deep percolation will reduce the salt load returned to the river (or groundwater) as well as reduce water loss. The adoption of the 'minimized leaching' concept of irrigation which reduces deep percolation should be of appreciable benefit for reducing salination of our water resources, especially in projects underlain by salt-laden sediments (van Schilfgaarde *et al.*, 1974). In addition, the interception of saline drainage water should likewise be beneficial. Intercepted saline drainage water can be desalted and reused, disposed of by pond evaporation or by injection into some isolated deep-aquifer, or it can be used as a water supply where use of brackish waters is appropriate.

Intercepting, Isolating, and Reusing Drainage for Irrigation

While there is an excellent opportunity to reduce the salt load contributed by drainage water through better irrigation management, especially through reductions in seepage and deep percolation, there are practical constraints which limit such reductions. However, the ultimate goal should be to maximize the utilization of an irrigation water supply in a single application with minimum drainage. To the extent that the drainage water still has value for use by a crop of higher salt tolerance, it should be used again for irrigation (Rhoades, 1977).

Drainage waters are often returned by diffuse flow or intentional direct discharge to the water course and automatically 'reused'. Dilution of return flow is often advocated for controlling water salinity. This concept has serious limitations when one considers the effect on the true volume of usable water, and it should not be advocated as a general method of salinity control (Rhoades, 1989). Diversions in excess of crop needs often provide return flows for irrigation downstream and help modulate the river flows. However, as already noted, such return is the mechanism by which much of the salt loading of rivers occurs, which, in turn, limits the kind of crops that can be grown. More significant is the fact that if the water being returned to the river is so saline that its use for crop production is nil, then dilution with purer water and using the mix for irrigation of crops of the same or lesser salt tolerance does not add to the usable water supply. One has, in this process of mixing, simply utilized the river as a combined 'delivery and disposal' system and mixed the usable and unusable waters into one blend, which is separated again by plants for their use. In an irrigated soil, the plant, through transpiration, 'distils' out the usable fraction of the mix (expending bioenergy to do so) and the 'unusable' fraction passes through the profile again and in the process displaces or 'picks up' more salt in its flow path. Greater flexibility for crop production results if the drainage water can be intercepted and isolated. Then the waters can be blended or used separately for irrigation or other uses. Once the drainage is mixed in surface waters, these alternatives are lost.

A recommended strategy for salinity control of river systems is to intercept drainage before it is returned to the river and to use it for irrigation by alternating it with the river water normally used during certain periods in the growing season of selected crops (Rhoades, 1984a). When the drainage water quality is such that its potential for reuse is exhausted, then it is discharged to evaporation ponds or other appropriate outlets. This strategy will conserve water, sustain crop production, and minimize the salt loading of rivers. It will also reduce the diversion of river water for irrigation. The feasibility of reusing drainage waters for irrigation is facilitated using the 'dual-rotation' management system of Rhoades (1984a, 1984b, 1984c). In this system, sensitive crops (lettuce, alfalfa, etc.) in the rotation are irrigated with 'low-salinity' river water, and salt-tolerant crops (cotton, sugar beet, wheat, etc.) are irrigated with drainage water. For the tolerant crops, the switch to drainage water is usually made after seedling establishment, preplant irrigations and initial irrigations being made with river water. The feasibility of this strategy is supported by the following (Rhoades, 1977; Rhoades *et al.*, 1988a, 1988b):

(a) The maximum possible soil salinity in the rootzone resulting from continuous use of drainage water does not occur when the water is used only for a fraction of the time.
(b) Substantial alleviation of salt build-up resulting from irrigation of salt-tolerant crops with drainage water occurs during the time salt-sensitive crops are irrigated with river water.
(c) Proper preplant irrigation and careful irrigation management during germination and seedling establishment leaches salts out of the seed area and from shallow soil depths.
(d) Data obtained in modelling studies and in field experiments support the credibility of this 'cyclic' reuse strategy.

Desalination of agricultural drainage waters for improving quality is not economically feasible even though it is to be implemented for the return flow of the Wellton–Mohawk project of Arizona. The high costs of the pretreatments, maintenance, and power are the deterrents. Only in extreme cases, or for political rather than technical reasons, is desalination advocated (van Schilfgaarde, 1979).

REFERENCES

Bernstein, L. (1974). Crop growth and salinity, in *Drainage for agriculture* (Ed. Jan van Schilfgaarde), *Agronomy*, **17**, 39–54.
Bingham, F. T., Strong, J. E., Rhoades, J. D. and Keren, R. (1985a). Effects of salinity and varying boron concentrations on boron uptake and yield of wheat, *Plant and Soil*, **97**, 345–51.
Bingham, F. T., Rhoades, J. D. and Keren, R. (1985b). An application of the Maas–Hoffman salinity response model for boron toxicity, *Soil Sci. Soc. Am. J.*, **49**, 672–4.
Bishop, A. A., Walker, W. R., Allen, N. L. and Poole, G. J. (1981). Furrow advance rates under surge flow systems, *J. Irrig. and Drainage Division, ASCE*, **107** (IR3), 257–64.
Corwin, D. L., Werle, J. W. and Rhoades, J. D. (1988). The use of computer aided mapping techniques to delineate potential areas of salinity development in soils: I. A conceptual introduction, *Hilgardia*, **56**(2), 1–17.
Corwin, D. L., Sorensen, M. and Rhoades, J. D. (1989). Field-testing of models which identify soils susceptible to salinity development, *Geoderma*, **45**, 31–64.
Dedrick, A. R., Replogle, J. A. and Erie, L. J. (1978). On-farm level-basin irrigation—save water and energy, *Civil Engineering*, **48**, 60–5.
Emerson, W. W. (1984). Soil structure in saline and sodic soils, in *Soil Salinity under Irrigation* (Eds. I. Shainberg and J. Shalhevet), Ch. 3.2, pp. 65–76, Springer Verlag, New York.
Ingvalson, R. D., Rhoades, J. D. and Page, A. L. (1976). Correlation of alfalfa yield with various indices of salinity, *Soil Sci.*, **122**, 145–53.
Jurinak, J. J. (1984). Thermodynamic aspects of the soil solution, in *Soil Salinity under Irrigation* (Eds. I. Shainberg and J. Shalhevet), Ch. 2.1, pp. 15–31, Springer Verlag, New York.
Jury, W. A. (1984). Field scale water and solute transport through unsaturated soils, in *Soil Salinity under Irrigation* (Eds. I. Shainberg and J. Shalhevet), Ch. 4.2, pp. 115–125, Springer Verlag, New York.
Keren, R. and Shainberg, I. (1984). Colloid properties of clay minerals in saline and sodic solution, in *Soil Salinity under Irrigation* (Eds. I. Shainberg and J. Shalhevet), Ch. 2.2, pp. 32–45, Springer Verlag, New York.
Keren, R., Bingham, F. T. and Rhoades, J. D. (1985a). Plant uptake of boron as affected by boron distribution between the liquid and the solid phases in soil, *Soil Sci. Soc. Am. J.*, **49**, 297–302.
Keren, R., Bingham, F. T. and Rhoades, J. D. (1985b). Effect of clay content in soil on boron uptake and yield of wheat, *Soil Sci. Soc. Am. J.*, **49**, 1466–70.
Law, J. P., Denit, J. D. and Skogerboe, G. V. (1972). The need for implementing irrigation return flow quality control, in *Proceedings of National Conference on Managing Irrigated Agriculture to Improve Water Quality*, pp. 1–17, Graphics Management Corp., Washington, D.C.
Maas, E. V. (1984a). Salt tolerance of plants, in *Handbook of Plant Science in Agriculture* (Ed. B. R. Christie), Vol. II, pp. 57–75, CRC Press Inc.
Maas, E. V. (1984b). Crop tolerance, *California Agriculture*, **38**, 20–1.

Maas, E. V. (1986). Salt tolerance of plants. *Appl. Agricultural Res.,* **1**, 12–26.

Maas, E. V. and Hoffman, G. J. (1977). Crop salt tolerance—current assessment, *J. Irrig. and Drainage Division, ASCE*, **103** (IR2), 115–34.

Maas, E. V. and Nieman, R. H. (1978). Physiology of plant tolerance to salinity, in *Crop Tolerance to Suboptimal Land Conditions* (Ed. G. A. Jung), Ch. 13, ASA Special Publication 32, pp. 277–299.

Oster, J. D. (1977). Various indices for evaluating the effective salinity and sodicity of irrigation waters, in *Proceedings Intl Salinity Conf., Texas Tech. University, Lubbock*, August, 1976, pp. 1–14.

Oster, J. D. and Rhoades, J. D. (1975). Calculated drainage water compositions and salt burdens resulting from irrigation with river waters in the western United States, *J. Environ. Qual.*, **4**, 73–9.

Rawlins, S. L. (1977). Uniform irrigation with a low-head bubbler system, *Agric. Water Mgmt*, **1**, 167–78.

Rhoades, J. D. (1972). Quality of water for irrigation, *Soil Sci.*, **113**, 277–84.

Rhoades, J. D. (1974). Drainage for salinity control, in *Drainage for Agriculture* (Ed. Jan van Schilfgaarde), Vol. 17, pp. 433–61, American Society of Agronomy, Madison, Wis.

Rhoades, J. D. (1977). Potential for using saline agricultural drainage waters for irrigation, in *Proceedings Water Mgmt for Irrigation and Drainage*, ASCE/Reno, Nevada, July 1977, pp. 85–116.

Rhoades, J. D. (1982). Reclamation and management of salt-affected soils after drainage, in *Proceedings of the First Annual Western Provincial Conf. Rationalization of Water and Soil Res. and Management*, Lethbridge, Alberta, Canada, 29 Nov.–2 Dec. 1982, pp. 123–97.

Rhoades, J. D. (1984a). Reusing saline drainage waters for irrigation: a strategy to reduce salt loading of rivers, in *Salinity in Watercourses and Reservoirs* (Ed. Richard H. French), Ch. 43, pp. 455–64, Butterworth, Stoneham, Mass.

Rhoades, J. D. (1984b). Using saline waters for irrigation, in *Proceedings Int'l Workshop on Salt Affected Soils of Latin America*, Maracay, Venezuela, 23–30 Oct. 1983.

Rhoades, J. D. (1984c). New strategy for using saline waters for irrigation, in *Proceedings ASCE Irrigation and Drainage Spec. Conf., Water Today and Tomorrow*, Flagstaff, Arizona, 24–26 July 1984, pp. 231–6.

Rhoades, J. D. (1988). The problem of salt in agriculture, *Yearbook of Science and the Future Encyclopedia Britannica*, Encyclopedia Britannica, Chicago, pp. 118–135.

Rhoades, J. D. (1989). Intercepting, isolating and reusing drainage waters for irrigation to conserve water and protect water quality, *Agr. Water Mgmt*, **16**, 37–52.

Rhoades, J. D. and Loveday, J. (1990). Salinity in irrigated agriculture, in *Irrigation of Agricultural Crops* (Eds. B. A. Stewart and D. R. Nielsen), ASA Monograph (in press).

Rhoades, J. D. and Merrill, S. D. (1976). Assessing the suitability of water for irrigation: theoretical and empirical approaches, *FAO Soils Bulletin*, **31**, 69–109.

Rhoades, J. D. and Suarez, D. K. (1977). Reducing water quality degradation through minimized leaching management, *Agric. Water Mgmt*, **1**, 127–42; Santa Barbara, Calif., Calif. Water Resources Ctr. Report 38, pp. 93–110.

Rhoades, J. D., Ingvalson, R. D., Tucker, J. M. and Clark, M. (1973). Salts in irrigation drainage waters. I. Effects of irrigation water composition, leaching fraction, and time of year on the salt compositions of irrigation and drainage waters, *Soil Sci. Soc. Am. Proceedings*, **37**, 770–4.

Rhoades, J. D., Oster, J. D., Ingvalson, R. D., Tucker, J. M. and Clark, M. (1974).

Minimizing the salt burdens of irrigation drainage waters, *J. Environ. Qual.*, **3**, 311–16.

Rhoades, J. D., Corwin, D. L. and Hoffman, G. J. (1981). Scheduling and controlling irrigations from measurements of soil electrical conductivity, in *Proceedings, ASAE, Irrigation Scheduling Conference*, Chicago, 14 Dec. 1981, pp. 106–15.

Rhoades, J. D., Bingham, F. T., Letey, J., Dedrick, A. R., Bean, M., Hoffman, G. J., Alves, W. J., Swain, R. V., Pacheco, P. G. and LeMert, R. D. (1988a). Reuse of drainage water for irrigation: results of Imperial Valley Study. I. Hypothesis, experimental procedures and cropping results, *Hilgardia*, **56**(5), 1–16.

Rhoades, J. D., Bingham, F. T., Letey, J., Pinter, P. J. Jr, LeMert, R. D., Alves, W. J., Hoffman, G. J., Replogle, J. A., Swain, R. V. and Pacheco, P. G. (1988b). Reuse of drainage water for irrigation: results of Imperial Valley Study. II. Soil salinity and water balance, *Hilgardia* **56**(5), 17–44.

Shainberg, I. (1984). The effect of electrolyte concentration on the hydraulic properties of sodic soils, in *Soil Salinity under Irrigation* (Eds. I. Shainberg and J. Shalhevet), Ch. 3.1, pp. 49–64, Springer Verlag, New York.

Shainberg, I. and Letey, J. (1983). Response of soils to sodic and saline conditions, *Hilgardia*, **52**, No. 2.

Suarez, D. L. (1981). Relationship between pH_c and SAR and an alternative method of estimating SAR of soil or drainage water, *Soil Sci. Soc. Am. J.*, **45**, 469–75.

Suarez, D. L. (1982). Graphical calculation of ion concentrations in $CaCO_3$ and/or gypsum soil solutions, *J. Environ. Qual.*, **11**, 302–8.

Szabolcs, I. (1985). Salt-affected soils—a world problem, *Int'l Symposium on Reclamation of Salt-Affected Soils*, China, 8–26 May 1985, pp. 30–47.

UN World Food Conference (1974). *Assessment of the World Food Situation: Present and Future*, Rome, 5–16 Nov. 1974.

United States Salinity Laboratory Staff (1954). Diagnosis and improvement of saline and alkali soils, US Department of Agriculture Handbook 60.

van Schilfgaarde, J. (Ed.) (1974). Drainage for agriculture, in *Agronomy*, Vol. 17, American Society of Agronomy, Madison, Wis.

van Schilfgaarde, J. (1976). Water management and salinity, *FAO Soils Bulletin*, **31**,. 53–67.

van Schilfgaarde, J. (1979). Water conservation potential in irrigated agriculture, in *Proceedings Soil Conservation Society of America's 34th Annual Mtg*, Ottawa, Canada, July–Aug. 1979.

van Schilfgaarde, J. (1984). Drainage design for salinity control, in *Soil Salinity under Irrigation* (Eds. I. Shainberg and J. Shalhevet), Ch. 6.2, pp. 190–7, Springer Verlag, New York.

van Schilfgaarde, J. and Rawlins, S. L. (1980). Water resources management in a growing society, in *Efficient Water Use in Crop Production* (Ed. T. R. Sinclair), Vol. 12, pp. 517–30, American Society of Agronomy, Madison, Wis.

van Schilfgaarde, J., Bernstein, L., Rhoades, J. D. and Rawlins, S. L. (1974). Irrigation management for salt control, *J. Irrig. and Drainage Division, ASCE*, **100** (IR3), 321–38; Closure: **102** (IR4), 467–9.

Wagenet, R. J. (1984). Salt and water movement in the soil profile, in *Soil Salinity under Irrigation* (Eds. I. Shainberg and J. Shalhevet), Ch. 4.1, pp. 100–14, Springer Verlag, New York.

Worstell, R. V. (1979). Selecting a buried gravity irrigation system, *Transactions of the American Society of Agricultural Engineering*, **22**(1), 110–14.

CHAPTER 5

Irrigation Development in Desert Environments

W. M. ADAMS and F. M. R. HUGHES

INTRODUCTION

Irrigation is the basis of most attempts to establish or enhance agricultural production in arid environments. Arid and semi-arid environments, with less than 500 mm of precipitation per year, account for somewhere between 30 and 40 million km^2 and over 14% of global population (Heathcote, 1983). Moreover, a substantial proportion of these arid lands lie within the poorest and least-developed countries, and a number of these, notably those that are members of the *Club du Sahel*, are predominantly arid or semi-arid. In these countries irrigated agriculture is widely held to have an important role to play. Increased agricultural production is necessary both to meet foodgrain demand and tackle growing problems of undernutrition and to contribute to national economic development.

Estimates of the incidence of chronic undernutrition vary because of the diversity of definition and the difficulty of measurement, but probably exceeded 535 billion in 1980 (17% of global population) (Grigg, 1985). The number has subsequently undoubtedly grown, despite the success of 'green revolution' technologies in Asia. In addition, shorter-term famine events, such as those in the Sahel and the Horn of Africa in the 1970s and 1980s, create crisis conditions on top of this base level of hunger. In Africa, 27 countries are identified by relief agencies as being at risk from drought and desertification (Curtis *et al.*, 1989). While overall average per capita consumption of calories in these countries has increased marginally, individual countries have experienced significant declines, for example Ghana (17%), Mali (16%) and Kenya (9%) (Curtis *et al.*, 1989).

The failure of food production to keep pace with population growth in many parts of sub-Saharan Africa through the 1970s was highlighted in the

Techniques for Desert Reclamation
Edited by A. S. Goudie
© 1990 John Wiley & Sons Ltd.

World Bank's appraisal in 1981, the Berg Report (World Bank, 1981), and both this and its successor stressed the need for increased agricultural production in terms of both the foodgrains self-sufficiency and export crop production. Irrigation is identified as an important factor in this development of agriculture within Africa, partly because of the tiny fraction of the continent irrigated (9 million ha, or 5% of the cropped land and only 0.3% of the total land area), about a fifth of that irrigated in India (FAO, 1987). Furthermore, 45% of the land area of Africa is too dry for crops and a further 15% is affected by variable rainfall conditions (FAO, 1987). Irrigation was a major feature of agricultural development policies in a number of sub-Saharan countries, for example Nigeria and Kenya, in the 1970s. Although these developments have proved both controversial and economically problematic (cf. Adams and Grove, 1984), there is no doubt that even in sub-Saharan Africa irrigation is now a fully accepted element in agricultural management.

THE NEED FOR IRRIGATION

Although the importance of nutrient limitation is increasingly being recognized (Ludwig, 1987; Breman and de Wit, 1983), plant growth in arid environments is primarily limited by water availability. This is a complex function of present and antecedent rainfall conditions, rates of evaporation (affected significantly by factors such as wind), and soil conditions (Jackson, 1977; Barrow, 1987). Plants of arid zones show adaptations to water shortfalls through features enhancing water acquisition or reducing water loss. Deep roots and high root/shoot ratios are common, as are small or specialized leaves, seasonal deciduousness and succulent life-form (McCleary, 1974; Cloudsley-Thompson, 1984). Arid zone crop plants such as sorghums and millets share some of these characteristics of drought resistance, and exhibit leaf curling under moisture stress and the capacity to survive substantial periods of drought.

A key factor in both wild plants and crops is their response to seasonality. In West Africa, Kowal and Kassam (1978) suggest that the growing season begins with 10 days of over 25 mm precipitation, followed by 2 weeks of rainfall exceeding evaporation. However, as Hulme (1987) points out in the context of the southern Sudan, growing seasons in arid areas are both difficult to define satisfactorily and highly variable. In the Sudan, the length of the dry season increases southwards, but there is also significant variability in the length and continuity of the wet season between years. It is the length and this variability of the growing season, which is typical of arid environments, which determines the need for and nature of irrigation.

Irrigation is often loosely defined. Fundamentally, it refers to the application of appropriate amounts of water to crop plants in a timely manner. As such, irrigation technology embraces both the application of water to crops and its removal in drainage. Irrigation therefore plays different roles in different agroclimatic zones. It is possible to distinguish between arid land irrigation, where rainfall is insufficient to support crop growth and agriculture is dependent on some form of irrigation, from supplementary irrigation in wetter areas, where irrigation can serve to both extend the growing season and insure against variability in wet-season length. In areas of high rainfall and low seasonality, engineering development may be chiefly aimed at reducing soil waterlogging and its attendant problems. Thus in West Africa, irrigation is necessary to support settled agriculture in the Sahel zone (with less than 300 mm precipitation); it allows double-cropping in the Sudano–Sahelian zone (with a unimodal rainy season and 700–1000 mm precipitation). Further south in the Sudano–Guinean savanna, irrigation is used to give higher yields and year-round cultivation. In the wettest tropical rain forest zone irrigation focuses on drainage (des Bouvrie and Rydzewski, 1977).

INDIGENOUS IRRIGATION

Contemporary irrigation projects tend to be large, costly and often centrally managed. In the Third World, such schemes often have a poor record in financial and socioeconomic terms. Partly for this reason, there has, in places like Africa, been increasing interest in alternative approaches to irrigation, particularly the notion of small-scale irrigation (Underhill, 1984; Moris *et al.*, 1984; Coward *et al.*, 1986; Adams and Carter, 1987; Adams and Anderson, 1988). Although in the technical engineering literature the distinction between rainfed and irrigated farming is fairly sharp, it is less useful in coping with small-scale and particularly locally organized or indigenous irrigation. Increasingly, therefore, the word irrigation is being defined more broadly to refer to a wide range of wetland cropping systems.

In many places, irrigation techniques predate the 'modern' postwar interest in irrigation as part of agricultural development in arid lands. Gibson (1974) describes the remarkable extent of irrigation in ancient and Islamic Mesopotamia, where a system of alternate-year fallowing evolved to cope with salinization. Similarly, irrigation was a key feature of agriculture on the coastal zone of northern Peru in the prehispanic era (Farrington, 1977; Farrington and Park, 1978), and the basis of civilization in Egypt (Butzer, 1976). The Roman origins of irrigation systems in eastern Spain are reviewed by Butzer *et al.* (1985).

Wetland areas have long been important in irrigated cultivation. Darch (1988), for example, discusses the drained field agriculture of prehispanic Latin America, and attempts to reconstruct Mayan drained fields. Analogous modern systems of cultivation on mounds and raised fields throughout the Tropics are discussed by Denevan and Turner (1974). The floodplain rivers of

Africa support a number of examples of cultivation based on flood-related cropping mechanisms (Adams and Anderson, 1988). Under these systems, water control is often slight, and irrigation involves a high degree of risk. Irrigated cropping represents part of a wider farming system, and a contribution to the spreading of the risks inherent in rainfed cultivation in areas of variable and fairly unpredictable rainfall (e.g. Adams, 1986). Choice of crops involves a complex mixture of ecological and socioeconomic decisions by farmers in response to the flooding regime of the wetland area, access to labour and availability of cultivable land.

In West Africa, rice (both Asian Rice (*Oryza sativa*) and West African rice (*Oryza glaberrima*)) is grown on both the rising (crue) and falling flood (décrue) in the Delta Intérieure of the Niger (Gallais, 1967; Gallais and Sidikou, 1978; Harlan and Pasquereau, 1969). Similar practices occur along the river Senegal (Watt, 1981; Boutillier and Schmitz, 1987), and in places on the Niger, and on the Benue, Sokoto and the Hadejia-Jama'are river systems (Adeniyi, 1973; Adams, 1986). Richards (1985, 1986) describes the swamp rice cultivation of inland Sierra Leone, and Linares (1981) the rice cultivation of the Diola of the Basse Casamance on the Senegalese coast, and similar coastal tide-related cultivation in Guinea and Guinea-Bissau. Sorghum is also important in residual soil moisture cropping on floodplain land, for example on the margins of Lake Chad, where 'masakwa' sorghum cultivation appears to be of considerable age (Connah, 1985), and the Sokoto Valley of northern Nigeria (Adams, 1986).

Many of the residual soil moisture crops cultivated in the wetlands of the Middle East and semi-arid West Africa depend on groundwater irrigation. Practices include simple hand-lifted calabash buckets from shallow unlined wells (for example in the Volta Delta; Chisholm and Grove, 1985) to shadoofs, which are widespread and long established in many areas. Animal-powered lifting devices are known from the Nile and the Middle East, for example the *saquia* of Persia, the *nora* of Portugal and the *cenia* of Spain, and appear to have been in use in the Roman Mediterranean, although perhaps subsequently spread by the Arabs (Butzer *et al.*, 1985). Although unusual in West Africa, oxen lift irrigation water in the Air Mountains of central Niger (Roger, 1984).

Rainwater harvesting techniques also have a long pedigree (cf. Gilbert-son, 1986). Such techniques were once well established, for example on the North African coast (Gale and Hunt, 1986), the Negev (Yair, 1983) and in the Yemen (Brunner and Haefner, 1986). Similar techniques are also known from places such as the Thar Desert (Kolarkar *et al.*, 1983a) and the Sonoran Desert (Nabhan, 1986). The notion has now been disseminated fairly widely in sub-Saharan Africa by aid agencies in the wake of the famines of the 1970s and 1980s, for example in Kenya and the West African Sahel (Bruins *et al.*, 1986).

The Middle East also supported systems of buried canals to distribute irrigation water, for example in Iran and Oman (Cressey, 1958; Beaumont, 1968; Wilkinson, 1974). Smaller irrigation pipes in rock, in some cases associated with terracing, have also been described from Palestine and the Yemen (Ron, 1985; Vogel, 1988). Hill irrigation systems, based on the diversion of flow from small mountain torrents, are widely distributed in the Himalayas, from Nepal, Pakistan and Afghanistan (e.g. Martin and Yoder, 1987; Butz, 1987; Jones, 1974; Howarth and Pant, 1987).

Similar irrigation systems, often called furrow irrigation systems, occur in various parts of the East African Rift Valley. The Konso of Ethiopia, the Marakwet and Pokot of Kenya and the Sonjo of Tanzania irrigate in this way, and it is clear that in the past there have been other examples of such irrigation, some of them extensive, notably that of Engaruka in Tanzania (Sutton, 1978, 1984). There are analogous irrigation systems in the Pare and Taita Hills, and most notably among the Chagga of Mount Kilimanjaro (Masao, 1974; Pike, 1965).

WATER SUPPLY FOR ARID LAND IRRIGATION

Irrigation can involve the development of a variety of different surface and groundwater sources. First, in arid areas, highly seasonal water flows are likely to demand seasonal or over-year storage, associated with a canal to carry water to the irrigation area, or the exploitation of remote all-year sources (such as snow-melt). Both long-distance water transfer schemes and storage-based schemes tend to be costly. Storage of seasonal flows in reservoirs can lead to significant environmental and socioeconomic impacts and costs. These are discussed below. In such cases, the economic returns from irrigation have to be considerable to justify the costs of project development. This aim is not always achieved, particularly in the Third World.

A series of large-scale irrigation schemes were developed in Nigeria in the 1970s based on storage of seasonal river flow. One of these is the Kano River Project, which lies south of Kano city, in northern Nigeria (Figure 5.1) (Wallace, 1979, 1980, 1981; Palmer-Jones, 1984). Feasibility studies began in 1969, and a pilot project in 1970. The target area is 42 000 ha, but by 1981 only about 12 500 ha (gross) had been completed (Adams and Hollis, 1988). Water is supplied from Tiga Dam on the Kano River, and transferred under gravity via a supply canal and an intermediate reservoir which holds one month's supply. The land was farmed prior to irrigation development, and the original farmers retain farms. These are mostly small (1–2 ha), although concentration of landholding is taking place as wealthier farmers acquire more land. A tractor hire service is available, and farmers coordinate production in large fields of 30 ha. The main crops are wheat, rice, tomatoes, millet and cowpeas. Yields are low, partly because of poor water-use efficiency, uncertainties about inputs and the availability of machine cultivation, and the attractions of alternative economic activities such as seasonal labour migration (Palmer-Jones, 1984). Uncertainties encourage farmers to grow crops such as long-season sorghum, which does not need wet season supplementary irrigation and which delays the planting date of dry-season irrigated wheat, thus contributing to low economic returns.

A second source of irrigation supply in arid areas is permanent surface water exploited using pumps and a canal system for delivery. Such schemes, for example the Bura Irrigation Development Scheme in Kenya described below (see Figure 5.1), demand large-scale pump installations and have high running costs. A key problem is that of the reliability of water supply, although there are other constraints such as the quality of water in times of low river flows, particularly where supply depends on the discharge from upstream irrigation areas (and which may therefore be saline or alkaline) or from urbanized areas (where it may be biologically polluted). A particularly acute example of the problem of unreliable water supplies is provided by the South Chad Irrigation Project in Nigeria.

The South Chad Irrigation Project draws water from Lake Chad through a 29 km canal, lifted by diesel pump to command the irrigation area, to the south-west of the lake, by gravity (Figure 5.1). Construction began in 1974, and the project was first commissioned in 1979. Stage 1 covers 32 000 ha (gross), but later stages were to have taken the total area irrigated up to 106 000 ha. In fact, very little land has been irrigated: only 7000 ha in 1983–4, and none in 1985–6 and 1986–7 (Kolawole, 1987). This failure is caused by the shrinkage of Lake Chad following the droughts of the 1970s and early 1980s. The lake began to fall below the intake level of the canal for part of the year in the late 1970s, and although the intake canal was extended over 20 km in 1985 and 1986, the lake continued to shrink. In 1985 and 1986 it failed to reach the minimum intake level, and the only agricultural production was by farmers who left the scheme to cultivate on the lake floor, sometimes using shallow hand-dug wells and petrol pumps (Kolawole, 1987, 1988).

The problems experienced with attempts to irrigate from Lake Chad are perhaps unusual, but in fact not entirely unpredictable. Rainfall variability has been high in the Sahel and Sudan zones on timescales of centuries and millennia, and the level of Lake Chad has fluctuated markedly even within the last century. Furthermore, earlier experience with irrigation in the area was marked by damage to schemes caused by *high* lake levels, which is evidence of the variability in such uncontrolled surface water resources (Kolawole, 1987; Palmer-Jones, 1987).

The third main source of water for irrigation in arid zones is groundwater, either from shallow aquifers in riverine floodplains (for example in Bangladesh or the valleys of northern Nigeria) or deeper aquifers. To an extent, pump schemes present less restrictions on the areas that may be irrigated, and

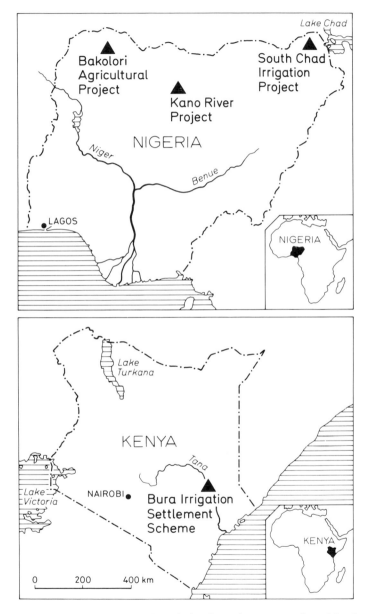

Figure 5.1   Location of major African irrigation schemes mentioned in the text

hence allow flexibility in the size, extent and location of irrigation. However, availability of power supply remains a key constraint on location. Electric pumps in particular demand rural power distribution, while petrol and diesel pumps present particular problems of maintenance, spares and fuel supply. The question of appropriate technology for rural water supply and irrigation has received considerable recent attention (e.g. Hofkes, 1981). There has been particular interest in the development of efficient human-powered lifting devices (O'Hea, 1983), solar pumps (Kenna and Gillet, 1985) and wind pumping (Abbot, 1982). The availability of water can also present problems for groundwater-based irrigation. The centre-pivot irrigation of the Kufrah Oasis in Libya, for example, depends of fossil water (*circa* 28 000 years old), tapped in the late 1960s at considerable cost (Allen, 1976). The exploitation of shallow aquifers through tubewells demands careful analysis of aquifer characteristics to avoid late-season shortages (Carter, 1984). Groundwater sources can also suffer from problems of quality, particularly salinity (e.g. Singh *et al.*, 1981). Shallow aquifers in association with human settlement can be contaminated by bacterial or viral pollution (Wellings, 1982).

THE APPLICATION OF IRRIGATION WATER

The extent of irrigation is limited by the availability of water, but an important secondary factor is the efficiency with which that water is used. Water-use efficiency embraces both the efficiency of the distribution network (from source to crop) and that within the field itself. If distribution canals are unlined, transmission losses can be considerable; Kinawy (1977), for example, estimates that the Ismailia Canal in Egypt loses 0.8 million m^3 annually because of its elevation and sandy section. This may not matter if supply is plentiful, canals are short and the slope and discharge are high, for example in the indigenous hill irrigation systems of Nepal. On flatter land or where water is short, transmission losses are more important, and the cost of lining, with puddled clay or even concrete, can be considerable. The increase in project costs of canal lining can be significant, and demand high productivity (and hence water-use efficiency) within the irrigated area.

Within irrigated fields, water can be applied by surface, sprinkler and trickle methods. Each has its merits and demerits (Anon, 1979). Surface application is the dominant method in the Third World. It involves the release of water by gravity onto the ground surface (perhaps by sluices at the field level or the use of flexible syphons from field ditches) either into broad basins or long furrows. In either case, land levelling to a fine level of tolerance is necessary to ensure that the water applied does not pond up and cause local waterlogging. Surface application may be undesirable in areas where the soil is either of markedly low infiltration capacity (where waterlogging may become a problem) or is sandy, where infiltration rates may be too high. There are unavoidable losses from evaporation as the water passes over the field surface. These can be significant in arid areas where humidity is low.

Unless the irrigation area is of low relief, surface application can be costly, and runs the risk of losing or burying topsoil in the land preparation process. Where land is already cultivated, for example in the Bakolori and Kano River Projects in northern Nigeria, land preparation for surface irrigation can be extremely disruptive and have serious socioeconomic impacts (Adams and Grove, 1984). Surface application can also be difficult to control, particularly if canals can be breached or drained by syphon, with consequent inefficiency of water use. On the other hand, construction, maintenance and operation all depend on relatively simple technology and resources that may, even within the rural Third World, be easily available.

Surface irrigation typically has a field irrigation efficiency (i.e. the ratio of water consumed by crops to that applied to the field) of about 40–50%. This is lower than that possible with sprinkler systems, which can approach 65%. In sprinkler systems, water is moved under pressure (usually pumped) and delivered through some kind of above-ground sprinkler nozzle. Typically, buried pipelines and field hydrants will be used, supplying lightweight movable pipes and sprinkler risers. Field sizes can be larger with sprinkler systems, with (theoretically) consequent gains from mechanization, and because water is applied at the crop plant rather than by moving across the land surface, applications are reduced. As a result, the chances of waterlogging are less, and larger areas can be irrigated from a given source of supply. Furthermore, land preparation is much reduced. On the negative side, sprinkler irrigation is costly to install and may involve foreign exchange costs to obtain sprinkler equipment. Furthermore, this equipment must be maintained and periodically replaced. Movement of pipes and risers is relatively labour intensive and the equipment may not be robust. There are also often extra demands for power to pump water to risers, again with associated costs. Sprinkler irrigation is obviously applicable in areas with soils of particularly high or low infiltration capacity, but not suitable where high winds are prevalent, since this greatly increases losses.

Many of the advantages of sprinkler irrigation are shared by the various forms of trickle irrigation. These involve piped delivery direct to the plant, thus minimizing evaporative losses within the field. The method is therefore particularly applicable to arid areas with low humidity and where water is in short supply. In some desert areas, trickle irrigation approaches a form of hydroponics, with both delivery system and growing medium integrated. The delivery system, often plastic pipe, is relatively costly, and trickle irrigation is therefore often associated with relatively high-value crops, such as vegetables or fruit, and capitalized production aimed at urban consumers or export. The poverty of many countries with extensive arid zones has led to a search for

cheap versions of trickle irrigation equipment, for example the use of discarded hospital infusion sets (Kolarkar *et al.*, 1983b).

ENVIRONMENTAL ASPECTS OF WATER STORAGE FOR IRRIGATION

Seasonal rainfall and interannual variations in precipitation in arid regions
often demand the storage of surface runoff for irrigation. Such impoundments, like those for flood control or hydroelectric power generation, bring
with them a series of significant environmental and socioeconomic costs.
Ecological impacts occur both in the area of the newly created reservoirs (e.g.
Ackerman *et al.*, 1973; Baxter, 1977) and in downstream areas.

Ecological change in impounded tropical reservoirs depends partly on the
chemistry, turbidity and temperature of inflowing waters and partly on the
nature of the substrate inundated. Fish communities and reservoir fisheries
may become established, often with an initial peak of population as nutrients
from flooded areas feed into the ecosystem, followed by a slump to a lower
level (Lowe-McConell, 1975; Welcomme, 1979; Davies, 1979).

The resettlement of evacuees from reservoir areas represents a significant
social and economic cost on storage-based irrigation development (Barrow,
1981; Gosling, 1979; Cernea, 1988; Goldsmith and Hildyard, 1984; Roggeri,
1985). The Bakolori Reservoir in northern Nigeria, for example, was built to
supply water to a 30 000 ha irrigation project. The dam was completed in
1978, flooding 12 000 people. Evacuees had no role in resettlement planning,
and there were significant disputes about the quality of resettlement villages
and substitute land, the survey of flooded property and the inadequacy of the
compensation that was paid (Wallace, 1980). Evacuees abandoned the resettlement villages in considerable numbers. Evacuees joined disgruntled
farmers from the irrigation scheme itself in blockades of the project area,
eventually dispersed with considerable violence by police (Adams, 1988).

Downstream impacts of dam construction depend on the nature of the
impact of the dam on the hydrological regime. This varies with the purpose
(or multiple purposes) of the dam. Dams for all-year irrigation store water at
seasons of high flow for use at times of low flow. Discharge beyond storage
capacity is usually spilled, allowing some flood flows to pass downstream in an
attenuated form. Such effects can be particularly significant where the river
regime is flashy and such peaks are common, for example in rivers in the
semi-arid tropics.

Reservoir siltation is a problem in river basins with high rates of sub-aerial
erosion and sediment transport, such as the Khashm el Girba Reservoir on
the Atbara River in Sudan, which supplies water to an irrigation scheme on
which Nubian evacuees from the Aswan Dam were resettled (Khogali, 1982).
By 1977 the original storage capacity had dropped by 59% through silt
accumulation, and it is estimated that it will drop by a further 38% by 1997

and that by 2025 there will be little water for irrigation. Irrigation efficiency and extent are already affected (Khogali, 1982). Possible solutions include more efficient irrigation water use and the construction of two upstream dams simply to act as silt traps. These might have additional power generation or irrigation functions, but obviously drastically alter the original economic justification for the project.

Sediment deposition and modification of the river regime lead to clear water releases below the dam and complex impacts on floodplain geo-morphology (e.g. Park, 1981; Petts, 1984). There can be a variety of eco-nomic impacts of such changes. It has been argued that the silt trapping effect of the Aswan dam in 1969 has had significant effects on the Nile Delta, has reduced soil fertility in the lower Nile and has allowed saline penetration of coastal aquifers (Biswas, 1980; Shalash, 1983). Lagoon and offshore fisheries have also been said to be affected (Kassas, 1973; George, 1973). The existence of erosion in the delta following the construction of the dam is confirmed by Murray *et al.* (1981). Bathymetric surveys offshore identified a new sand ridge system acting as a sink for eroded material.

Adverse impacts of dams on downstream fisheries have been reported from many tropical river systems (e.g. Davies, 1979; Lowe-McConell, 1975). Im-pacts on floodplain agriculture can also be significant. For example, the Bakolori Dam on the Sokoto River in Nigeria brought about a significant reduction in peak flows in the Sokoto (Adams, 1985). This in turn reduced the depth, duration and extent of inundation in the floodplain for 120 km down-stream. The area of floodplain cultivated in the wet season fell as a result, with a particularly marked fall in the area under rice (Adams, 1985). Dry season cropping also shrank in area, and irrigation with simple wells, which was already widely practised, became more costly and more specialized because of falls in the water table (Adams, 1985).

Environmental Aspects of Arid Land Irrigation

The particular environmental impacts of water storage for irrigation have been discussed above. However, a range of other environmental implications of irrigation development present significant constraints on development under some conditions (Diamant, 1980). One of the most important is that of the desertification of soil through the concentration of salt or soda (Warren and Maizels, 1977; Warren and Agnew, 1988). The problem of salinity is discussed in detail by Rhoades (Chapter 4, this volume) and need not be considered in detail here. However, it is important to note that the problem is widespread and a significant limitation on the effectiveness and effective extent of irrigation in many arid regions. Up to 50% of irrigated land may be affected by alkalinity, salinity or compaction (Kovda, 1977; Goldsmith and Hildyard, 1984).

The problems of salinity and alkalinity are particularly severe in countries such as Egypt (El-Gabaly, 1979; Ibrahim, 1982), Pakistan (Bokhari, 1980) and Iraq (Saad, 1982), but also afflict irrigation in industrialized countries, such as the USA (Law and Hornsby, 1982) and Australia (Thomas and Jakeman, 1985). Particular problems include the use of water naturally high in salts or alkali, re-use of return flows from irrigation schemes for further downstream irrigation (Urroz, 1977), the existence of shallow water tables in areas of high evaporative demand, poor drainage and the lack of provision of drainage in scheme development, and management problems of overwatering. Often these problems occur together, for example in the Murray Basin in Australia (Pels and Stannard, 1977) or in the Grand Valley, USA (Skogerboe *et al.*, 1982).

There are technical solutions to problems of salinity and alkalinity. At the scheme scale these include lining conveyancing canals and installing drainage systems, while at the field scale, the installation of surface, shallow soil or deep tubewell/pump drainage, lining secondary and tertiary canals and improving water use through a change in application method or reform of the relations between farmer and scheme management (Skogerkoe *et al.*, 1982; Pels and Stannard, 1977). There is also increasing interest in the breeding of salt-tolerant crops (e.g. Charnock, 1988). Alternative strategies include the separation of saline discharges so that they do not reenter rivers used downstream, as Dukhovny and Litvak (1977) describe for the Syr Darya in central Asia, and desalination, which has been used to deal with the saline discharge of the Welton–Mohawk sytem in Arizona (Schilfgaarde, 1982). While technically feasible, many strategies of this kind are expensive, and may not be cost effective. Nonetheless, the high capital cost of irrigation schemes means that further investment in rehabilitation may seem a viable and necessary strategy.

Water quality is not only a problem for irrigation itself, but for other downstream uses. Mineral quality, temperature, taste and turbidity can all be affected by irrigation (Hotes and Pearson, 1977). There can also be problems of biological purity and enhanced levels of nitrates, phosphates and agricultural pesticides. These can seriously limit the potential of downstream use for purposes such as drinking. Irrigation consumption can also have a significant impact on the quantity of water reaching downstream areas, as the falling levels of closed basin lakes such as the Caspian and Aral Seas demonstrate (Hollis, 1978).

Pesticides, particularly insecticides, are increasingly seen as an essential element in irrigated cropping, partly because of the need to obtain the guaranteed high yields necessary to repay the capital cost of scheme development. Crop losses through pest attack are a major problem in the Third World, both during crop growth and during storage. It is estimated, for example, that in-field crop losses are 41.6% in Africa—13% from insects,

12.9% from disease and 15.7% from weeds (Ghatak and Turner, 1978). Losses in Asia are 43.3%, in Latin America 33%. Nonetheless, there are problems of operator safety, the impact of pesticides on non-target organisms and the development of resistance of target organisms.

The problems of pest resistance to pesticides under continuous use is well documented (Bull, 1982), the most commonly quoted example being the growing resistance of mosquitos to organochlorine pesticides used in malaria control. There is also evidence that the use of organophosphorus and carbamate insecticides in agriculture is implicated in some instances in the development of pesticide resistance in mosquito vectors of malaria (PEEM, 1987). In Malaysia the number of rice pests resistant to at least one pesticide rose from 8 in 1965 to 14 in 1975 (Bull, 1982). It is now clear, for example, that the brown plant hopper (*Niliparvata ingens*) is controlled by a wide guild of predators, and rapid resurgence follows accidental control by pesticides of these organisms. There is now an FAO/UNEP Cooperative Programme on Integrated Pest Control in South East Asia tackling this problem, trying to reduce levels of pesticide use and devise integrated approaches to pest control including breeding rice varieties with pest-resistant attributes (PEEM, 1987).

Similar problems emerged with cotton spraying against whitefly on the Gezira Scheme in the Sudan. The area sprayed increased rapidly from 600 ha in 1946 (0.7% of the total area) to 98 000 ha (100%) in 1954. The number of sprays then increased from 1 per year in the late 1950s to 9 per year in the late 1970s. By 1976, 2500 tonnes of insecticide were being used per year at Gezira, yet despite this effort and the massive rise in production costs it represented, pest losses continued (Bull, 1982).

With growing pesticide use in the Third World risks to operators are also growing. High standards of safety may be specified by pesticide manufacturers, but Third World farmers are likely to be unfamiliar with the nature of the chemicals they are using, perhaps unable to read health warnings and unable to acquire necessary protective clothing. It is estimated that there are 375 000 cases of poisoning by pesticides in the Third World, 10 000 of them fatal (Caufield, 1984). In Vavuniya in north-east Sri Lanka, for example, there were 938 deaths from pesticide poisoning in 1977, more than malaria, tetanus, diphtheria, whooping cough and polio put together (Bull, 1982, p. 44).

The ecological changes associated with irrigation also have significant direct effects on both pest and disease organisms. Ali (1977), for example, describes the increased density of the cotton leafworm (*Spodoptera littoralis*) in irrigated areas in Egypt following the development of irrigation allowed by the construction of the Aswan High Dam. More familiar, perhaps, are the problems of the expansion of schistosomiasis (bilharzia) and malaria with irrigation development (Betterton, 1984; Farid, 1977; Hill *et al.*, 1977). Schistosomiasis in particular is associated with irrigation, the snails which are the intermediate hosts of the parasite thriving in slow-moving water in irrigation canals, particularly those which are badly maintained and thick with weed (cf. Mitchell, 1977). Amin (1977) suggests that 60% of the population (and 80% of children) on the Gezira scheme suffer from schistosomiasis. Control is conventionally through chemical molluscicides, although better management and (on new schemes) appropriate engineering design of canals can help.

THE BURA IRRIGATION SETTLEMENT PROJECT: A CASE STUDY

Introduction

The Bura Irrigation Settlement Project in Kenya is one of a series of large-scale irrigation schemes initiated in sub-Saharan Africa in the 1970s. Construction began in 1979 on a scheme of 6700 ha, on the West bank of the Tana River, to grow cotton using settler tenants from all over Kenya (Figure 5.2). This was expected to be the first phase of a larger development on both banks of the river, and supplied by gravity from a barrage on the Tana at Nanighi. The final area was to be 12 500 ha on the west bank and 25 000 ha on the east bank, making Bura one of the largest irrigation schemes in sub-Saharan Africa outside the Sudan. Cotton was seen as an important export crop and foreign exchange earner.

It was envisaged that there would be 5150 tenant families on the first 6700 ha, but by 1984 there were only 1800 tenants, cultivating only 1100 ha of maize and 2050 ha of cotton. Major problems had been revealed with the quality of soils, water supply was unreliable and yields and returns were low. The scheme was experiencing severe financial and management problems. The story of the failure of Bura is complex, but it provides an important cautionary tale to those who see large-scale irrigation as an automatic and unproblematic solution to the problems of impoverished and drought-prone arid environments in the Third World.

History of Irrigation at Bura

The notion of irrigation on the Tana River has a long history. In 1948, two 40 000 ha irrigation schemes were proposed between Bura and Grand Falls in the Upper Tana, but subsequently abandoned because of remoteness and poor soils (World Bank, 1985). Between 1953 and 1957, a short-lived rice project existed on the East bank opposite Hola. At Hola itself, a Pilot Irrigation Scheme was built in the 1950s. Studies of irrigation potential in the Lower Tana, carried out between 1963 and 1967, investigated the possibility of two large-scale irrigation schemes of 100 000 and 120 000 ha following upstream dam constuction for the generation of hydropower. Although

149

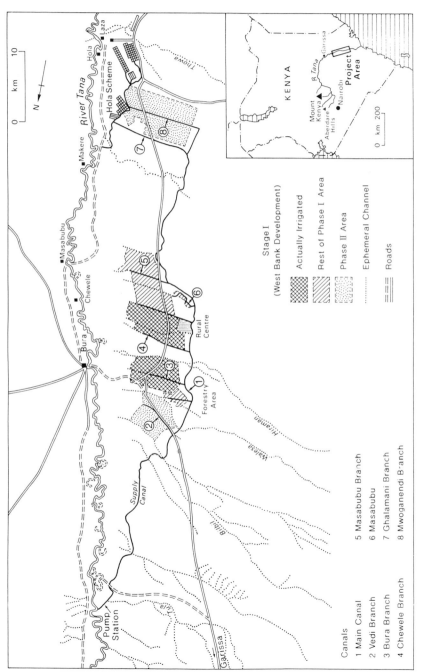

Figure 5.2 Layout of the Bura Irrigation Settlement Project, Kenya

technically feasible, this was thought to be economically unattractive (Acres/
ILACO, 1967).

Subsequent feasibility studies in the 1970s identified smaller areas of
potentially irrigable land, and a specific feasibility study was done on an area
of 4000 ha (subsequently raised to 14 000 ha) on the west bank of the river
near Bura (ILACO, 1973, 1975). At this stage, poor soil quality was considered a sufficiently major constraint that only economies of scale could
compensate for it, with water supply by gravity from a fixed weir. Subsequently, donors (including the World Bank) were found for this scheme. The
details of financing and project planning were complicated, and are reviewed
in World Bank (1985) and Adams (in press).

Predicted Problems

At the feasibility study stage, it was predicted that poor soil quality would be
a major constraint on yields: soils were generally too shallow and alkaline and/
or saline. A shortage of fuelwood was also predicted, because the arrival of
settlers and their families would more than double the existing local population of Orma pastoralists and Malekote and Pokomo riverine cultivators.
The final design for Bura included a 4500 ha irrigated fuelwood plantation for
the first 6700 ha of the scheme (MacDonald, 1977). Provision of fuelwood was
considered important not only because tenants would have great difficulty in
collecting it, but also because of impacts on the riverine forest strip, which
was of known conservation importance because of high species diversity and
the presence of two species of endemic primates.

Problems in Implementation

Bura was dogged with financial and other problems from the feasibility and
planning stages (Adams, in press). When construction of Bura began in mid
1979, two years behind schedule, the scheme was seriously underfunded.
Costs had risen to 187% of estimates in the World Bank's Appraisal Report
(World Bank, 1985). More importantly, perhaps, the proportion of total costs
to be borne by the Kenyan Government had risen to 50%, both because of
the additional capital costs of the new design and the withdrawal of one
donor. An Interministerial Committee formed in 1982 recommended cost
cutting: a reduction in the irrigated area to 3900 ha, cancellation of the
barrage, cotton ginnery and 132 kV transmission line.

There have been a number of other environmental and socioeconomic
problems at Bura. Dissatisfaction among tenant families has arisen because of
poor housing, educational and health facilities, and because small farm plots
(1.25 ha), low incomes and the credit system have pushed farmers into debt.
Irrigation water supply has been erratic largely because of the failure of the

pumps which lift water from the Tana River into the supply canal (due primarily to problems of maintenance, spare parts and fuel supply), exacerbated by problems of canal siltation. Within fields, irrigation efficiency is low, partly because of varying furrow slopes, partly because of the low infiltration capacity of soils. Water supply irregularities have contributed to poor yields of both cotton and maize, and poor settler incomes. In 1985, for example, production costs (fertilizer, water charges, aerial spraying) were estimated to be over 9000 K.Sh, while net incomes ranged between 3791 and 7942 K.Sh. (Vainio-Mattila, 1987).

Scheme management has been a constant source of problems at Bura. The National Irrigation Board (NIB) was a Nairobi-based organization with no previous experience with irrigation on this scale. In June 1985, following the critical World Bank Mid-Term Evaluation (World Bank, 1985), Bura was given autonomy under the Ministry of Agriculture and Livestock Development, with management on site. In January 1986, the Kenyan President made a surprise visit to the scheme, and set up a third managerial system with a greater role for the provincial and district administration. In 1986, the World Bank withdrew from further involvement with the scheme (Vainio-Mattila, 1987; Moris and Thom, 1987).

However, Moris and Thom (1986) argue that Bura would have encountered severe operation difficulties even with excellent management. First, they point out that the reduced size of the scheme (3900 ha, 3000 tenants) means that even with good yields the scheme would incur annual deficits on agricultural operations and social services. Second, reliance on pumps remote from the management centre creates high risks of breakdown and interrupted water supply. Third, good yields demand timely planting, and hence highly coordinated operation of the tractor service, thus placing further stresses on the management system. Furthermore, poor roads and black cotton soils make access and field operations difficult after rainfall. Fourth, the performance of both public and private agencies at Bura has been poor, for example in the provision of health facilities. Fifth, tenants have been impoverished by the system of charging for water and agricultural services regardless of the quality of that service.

Gitonga (1985) suggests that farmer incomes could be enhanced by the improvement of crop yields though the encouragement of better agronomic practices, increasing the size of tenants' plots, the adoption of high-yielding varieties and new cropping patterns, and the enhancement of fixed crop prices.

Impacts on Bura on the Lower Tana

Conventional economic analysis of Bura will inevitably reflect its slow implementation, high per hectare development costs, high running costs and low economic returns. However, such analysis must also take into account the costs of less obvious socioeconomic and environmental impacts on the Lower Tana. Although the planning of Bura was based on the assumption that the project area was uninhabited, it was in fact used by Orma pastoralists. They have lost grazing land and have been barred from access to dry season water in the river by the 40 km supply canal (which runs roughly parallel to the river) and irrigated fields. Attempts to provide drinking points alongside the canal, filled when the canal is full, have proved ineffective. A small number of Orma have settled on the scheme and many others have become partially sedentary in the area because of the availability of water and informal employment. This has caused severe overgrazing of the surrounding area (Ledec, 1987).

There has been a major environmental impact on the evergreen forests of the Tana River floodplain (Hughes, 1984, 1987, 1988). These forests are relatively species-rich forests and support endemic primates, the Tana River Red Colobus and the Tana Mangabey (Marsh, 1976; Homewood, 1976). They merit protection for these reasons, and also because of their role in channel stabilization. Their ecology and composition are discussed in Marsh (1976) and Hughes (1984, 1988, 1989). The forests have been under threat at Bura since development began because of the demands for fuelwood by the construction workforce, amounting to several thousand people. Tenants began to arrive in 1981, and also required fuelwood. However, the proposed fuelwood plantation was axed as an economy measure. The first tenants initially used stockpiles of cleared acacia scrub for fuel, then began to deplete the surrounding bush. It did not take long for the Malekote, resident in the floodplain forest strip, to identify the market for charcoal, and at an early stage they began to convert *Acacia elatior* from the forest into charcoal for sale to Bura residents, particularly administrative staff (Hughes, 1987).

In 1985, the World Bank emphasized that the irrigated fuelwood plantation must be initiated both to reduce the stress of increasingly arduous fuelwood collection on tenants and to protect the natural vegetation of the area (World Bank, 1985). By 1988, the population of the scheme was about 13 000 people, with an additional 3000–4000 administrative and business workers (Johansson, 1988). Eventually, FINNIDA funded the Bura Fuelwood Plantation Project (BFPP), whose main objective was to provide fuelwood for the scheme population through 600 ha of irrigated plantation.

Research trials began in 1984. By 1987, a number of recommendations were being made, including fairly detailed planting plans for the forestry area itself, planting in all areas of seepage or drainage water overflow, and water harvesting (Johansson, 1988). Emphasis was placed on unirrigated planting because of the problem of the uncertainty of water supply. Another aspect of the BFPP has involved a stove programme in tenant homes, resulting in only small savings in fuelwood use, but positive lifestyle changes (Vainio-Mattila,

1987). It is estimated that unless these proposals are implemented, large areas of the floodplain forest (particularly those where acacias dominate) will be destroyed within 5–10 years (Hughes, 1987).

The construction of Bura involved the clearance of large areas of acacia bush, with resulting loss of wildlife habitat. The 40 km supply canal also represents a barrier to wildlife seeking to move between the bush and the river. Ledec (1987) argues that a more significant impact of bush clearance is increased wind erosion. This problem is exacerbated by the fact that only 3900 ha of the 6700 ha cleared will now be irrigated. Dust-storms lead to problems of siltation in small irrigation channels. Wildlife poaching is perceived to be a problem in the vicinity of Bura, although there are difficulties also with the loss of crops to wildlife damage, 180 ha of maize being lost in 1983 (Ledec, 1987).

A final environmental impact of the Bura scheme is that of pesticides. A number of chemicals, including the organochlorine Thiodon are used to control various cotton pests, and aerial spraying methods are used. The impacts of this spraying on tenant and operator health, on non-target organisms within the scheme and through spraydrift on areas outside the scheme may be significant, but have not been assessed.

CONCLUSION

Although the design of irrigation systems is highly technical, it is becoming clear that the key determinant of the effectiveness of irrigation, certainly in the Third World, is management (Bottrall, 1985; Carruthers, 1981a; Chambers, 1988). Indeed, Carruthers (1981b) highlights the growing unease among irrigation experts and others about the gap between potential and actual performance of irrigation schemes. There are a variety of conclusions often drawn. Certainly operation and maintenance ('O and M') are too often ignored (e.g. Jayaraman, 1982). Certainly, too, it is necessary to extend attempts to understand irrigation schemes beyond the conventional technical fields of agronomy and water control into questions of management systems.

Beyond this, it is clear that issues of political economy, of the working of bureaucracies in the real (and corrupt) world (cf. Wade 1982) and the pressures on tenant farmers are vital to an adequate understanding of irrigation schemes and the conditions under which they realize the aims of their designers. Perhaps Carruthers (1981a) is right to suggest that the greatest potential for increasing agricultural production lies not with the construction of new irrigation schemes but with the reorganization, rehabilitation and improvement of existing irrigated areas.

REFERENCES

Abbot, V. (1982). Windpumps bring water to the Turkana desert, *Waterlines*, **12**(2), 6–7.

Ackerman, W. C., White, G. F. and Worthington, E. B. (Eds.) (1973). Man-made lakes: their problems and environmental effects, American Geophysical Union, Monograph 17.

Adams, W. M. (1985). The downstream impacts of dam construction: a case study from Nigeria, *Trans. Inst. Brit. Geogr. NS*, **10**(3), 292–302.

Adams, W. M. (1986). Traditional irrigation and water use in the Sokoto Valley, Nigeria, *Geog. Journal*, **152**, 30–43.

Adams, W. M. (1988). Rural protest, land policy and the planning process on the Bakolori project, Nigeria, *Africa*, **58**, 315–336.

Adams, W. M. (in press). How beautiful is small? Scale, control and success in Kenya irrigation. *World Development* (forthcoming).

Adams, W. M. and Anderson, D. M. (1989). Irrigation before development: indigenous and induced change in agricultural water management in East Africa, *African Affairs*, **87**, 519–35.

Adams, W. M. and Carter, R. C. (1987). Small scale irrigation in Sub-Saharan Africa, *Progress in Physical Geography*, **11**, 1–27.

Adams, W. M. and Grove, A. T. (Eds.) (1984). *Irrigation in Tropical Africa: Problems and Problem-solving*, Cambridge, African Studies Centre Monograph No. 3, 1984.

Adams, W. M. and Hollis, G. E. (1989). *Hydrology and Utilisation of a Sahelian Floodplain Wetland*, Hadejia Nguru Wetlands Conservation Project, Sandy, Beds.

Adeniyi, E. O. (1973). Downstream impact of the Kaini Dam, in *Kainji: A Nigerian Man-made Lake: Socio-economic Conditions* (Ed. A. L. Mabogunje), Nigerian Institute for Social and Economic Research, Ibadan.

Ali, A. M. (1977). Impact of changing irrigation on agricultural pests and wildlife in Egypt, in *Arid Land Irrigation in Developing Countries: Environmental Problems and Effects* (Ed. E. B. Worthington), pp. 331–2, Pergamon Press, Oxford.

Allen, K. A. (1976). The Kufrah agricultural scheme, *Geog. Journal*, **142**, 50–6.

Amin, M. A. (1977). Problems and effects of schistosomiasis in irrigation schemes in the Sudan, in *Arid Land Irrigation in Developing Countries: Environmental Problems and Effects* (Ed. E. B. Worthington), pp. 407–11, Pergamon Press, Oxford.

Anon (1979). Irrigation: a comparison between sprinkler and surface, *West African Farming*, March **1979**, 27–33.

Barrow, C. J. (1981). Health and resettlement consequences and opportunities created as a result of river impoundment in developing countries, *Water Supply and Management*, **5**, 135–50.

Barrow, C. J. (1987). *Water Resources and Agricultural Development in the Tropics*, Longman, London.

Baxter, G. (1977). Environmental effects of dams and impoundments, *Ann. Rev. Ecol. Syst.*, **8**, 255–83.

Beaumont, P. (1968). Quanats on the Varamin Plain, Iran, *Trans, Inst. Br. Geogr.*, **45**, 169–80.

Betterton, C. (1984). Ecological studies on the snail hosts of schistosomiasis in the South Chad Irrigation Project Area, Borno State, northern Nigeria, *J. Arid Environments*, **7**, 43–58.

Biswas, A. K. (Ed.) (1980). The Nile and its environment, *Water Supply and Management*, **4**, 1–113.

Bokhari, S. M. H. (1980). Case study on waterlogging and salinity problems in Pakistan, *Water Supply and Management*, **4**, 171–92.

Bottrall, A. (1985). Managing large irrigation schemes: a problem of political economy, ODI Occasional Paper 5.

Boutillier, J. and Schmitz, J. (1987). Gestion traditionel des terres (système de décrue/système pluvial) et transition vers l'irrigation, *Cah. Sci. Hum.*, **23**(3–4), 533–54.

Breman, H. and de Wit, C. T. (1983). Rangeland productivity and exploitation in the Sahel, *Science*, **221**(4618), 1341–7.

Bruins, H. J., Evenari, M. and Nessler, U. (1986). Rainwater-harvesting agriculture for food production in arid zones: the challenge of the African famine', *Applied Geography*, **6**(1), 13–32.

Brunner, U. and Haefner, H. (1986). The successful floodwater farming systems of the Sabeans, Yemen Arab Republic, *Applied Geography*, **6**(1), 77–78.

Bull, D. (1982). *A Growing Problem: Pesticides and the Third World Poor*. Oxfam Books, Oxford.

Butz, D. A. O. (1987). Irrigation agriculture in high mountain communities: the example of Hopar, Pakistan, M.Sc. Thesis, Wilfred Laurier University.

Butzer, K. W. (1976). *Early Hydraulic Civilisation in Egypt: A Study in Cultral Ecology*, University of Chicago Press.

Butzer, K. W., Mateu, J. F., Butzer, E. K. and Kraus, P. (1985). Irrigation agro-systems in Eastern Spain. Roman or Islamic origins?, *Annals of the Association of American Geographers*, **75**, 479–509.

Carruthers, I. (1981a). Do engineers make good managers?, *Water Supply and Management*, **5**, 1–3.

Carruthers, I. (1981b). Neglect of O and M in irrigation: the need for new sources and forms of support, *Water Supply and Management*, **5**, 53–65.

Carter, R. C. (1984). Groundwater retrieval and delivery for small scale irrigation in northern Nigeria, in *Report of the Second Fadama Seminar* (Ed. N. P. Chapman), pp. 19–35, Bauchi State Agricultural Development Project, Azare, Nigeria.

Caufield, C. (1984). Pesticides: exporting death, *New Scientist*, 16 August 1984, pp. 15–17.

Cernea, M. (1988). Involuntary resettlement in development projects: policy guidelines in World Bank-financed projects, *World Bank Technical Paper 80*, Washington.

Chambers, R. (1988). *Managing Canal Irrigation: Practical Analysis from South Asia*, Cambridge University Press.

Charnock, A. (1988). Plants with a taste for salt, *New Scientist*, 3 December 1988.

Chisholm, N. G. and Grove, J. M. (1985). The Lower Volta, in *The Niger and Its Neighbours: Environmental History and Hydrobiology. Human Use and Health Hazards of the Major African Rivers* (Ed. A. T. Grove), Balkema, Rotterdam.

Cloudsley-Thompson, J. L. (Ed.) (1984). *Sahara Desert*, Pergamon Press, Oxford.

Connah, G. (1985). Agricultural intensification and sedentarism in the firki of northeast Nigeria, *Prehistoric intensive agriculture in the Tropics* (Ed. I. S. Farrington), BAR International Series 232, London, pp. 785–6.

Coward, E. W., McConner, R. J., Ramchand, O., Ssennyonga, J., Arao, L. and Gichuki, F. (1986). Watering the Shamba: current public and private sector activities for small-scale irrigation development in Kenya, Water Management Synthesis II Project Report 40.

Cressey, G. B. (1958). Quanats, kargey and foggaras, *Geogrl. Review*, **48**, 27–44.

Curtis, D., Hubbard, M. and Shepherd, A. (1989). *Preventing Famine: Policies and Prospects for Africa*, Routledge, London.

Darch, J. P. (1988). Drained field agriculture in tropical Latin America: parallels from past to present, *J. Biogeography*, **15**, 87–95.

Davies, B. R. (1979). Stream regulation in Africa: a review, in *The Ecology of Regulated Streams* (Eds. J. V. Ward and J. A. Stanford), Plenum Press, New York.

Denevan, W. M. and Turner, B. L. II (1974). Forms, functions and associations of raised fields in the Old World Tropics, *J. Trop. Geogr.*, **39**, 24–33.

des Bouvrie, C. and Rydzewski, J. R. (1977). Irrigation, in *Food Crops of the Lowland Tropics* (Eds. C. L. A. Leakey and J. B. Wills), pp. 161–94, Oxford University Press, Oxford.

Diamant, B. Z. (1980). Environmental repercussions of irrigation development in hot climates, *Environmental Conservation*, **7**, 57–8.

Dukhovny, V. and Litvak, L. (1977). Effect of irrigation on Syr Darya water regime and water quality, in *Arid Land Irrigation in Developing Countries: Environmental Problems and Effects* (Ed. E. B. Worthington), pp. 265–75, Pergamon Press, Oxford.

El-Gabaly, M. M. (1979). Secondary salinisation and sodication in Egypt: a case study, *Water Supply and Management*, **3**, 174–200.

FAO (1987). *Consultation on Irrigation in Africa, Proceedings of the Consultation on Irrigation in Africa*, Lomé, Togo, 21–25 April 1986.

Farid, M. A. (1977). Irrigation and malaria in arid lands, in *Arid Land Irrigation in Developing Countries: Environmental Problems and Effects* (Ed. E. B. Worthington), pp. 413–19, Pergamon Press, Oxford.

Farrington, I. S. (1977). Land use, irrigation and society on the North Coast of Peru in the prehispanic era, *Zeitschrift für Bewassurungswirtschaft*, **12**, 151–86.

Farrington, I. S. and Park, C. C. (1978). Hydraulic engineering and irrigated agriculture in the Moche Valley, Peru: c.AD 1250–1532, *J. Arch. Sci.*, **5**, 255–68.

Gale, S. J. and Hunt, C. O. (1986). The hydrological characteristics of a floodwater farming system, *Applied Geography*, **6**, 33–42.

Gallais, J. (1967). *La Delta Intérieur du Niger: Etude de Géographie Regionale*, Institute Fondamental d'Afrique Noir, Dakar.

Gallais, J. and Sidikou, A. H. (1978). Traditional strategies, modern decision-making and management of natural resources in the Sudan Sahel, Man and the Biosphere technical note, UNESCO, Paris.

George, C. J. (1973). The role of the Aswan High Dam in changing the fisheries of the southeastern Mediterranean, in *The Careless Technology: Ecology and International Development* (Eds. M. T. Farvar and J. P. Milton, pp. 159–78, Stacey, London.

Ghatak, S. and Turner, R. K. (1978). Pesticide use in less developed countries: economic and environmental considerations, *Food Policy*, **3**, 136–46.

Gibson, M. (1974). Violation of fallow and engineered disaster in Mesopotamian civilisation, in *Irrigation's Impact on Society* (Eds. T. E. Downing and M. Gibson), pp. 7–19, University of Arizona Press, Tucson.

Gilbertson, D. D. (Ed.) (1986). Runoff farming in arid lands, *Applied Geography*, **6**(1).

Gitonga, S. M. (1985). Bura Irrigation Settlement Project, unpublished manuscript.

Goldsmith, E. and Hildyard, N. (1984). *Social and Environmental Impacts of Large Dams*, Vol. 1, Ecologist Magazine, Wadebridge, Cornwall.

Grigg, D. B. (1985). *The World Food Problem 1950–1980*, Basil Blackwell, Oxford.

Harlan, J. R. and Pasquereau, J. (1969). Décrue agriculture in Mali, *Economic Botany*, **23**, 70–4.

Heathcote, R. L. (1983). *The Arid Lands: Their Use and Abuse*, Longman, London.

Hill, M. N., Chandler, J. A. and Highton, R. B. (1977). A comparison of mosquito populations in irrigated and non-irrigated areas of the Kano Plains, Nyanza Province, Kenya, in *Arid Land Irrigation in Developing Countries: Environmental Problems and Effects* (Ed. E. B. Worthington), pp. 307–15, Pergamon Press, Oxford.

Hofkes, E. H. (Ed.) (1981). *Small Community Water Supplies: Technology of Small Water Supply Systems in Developing Countries*, Wiley, Chichester.

Hollis, G. E. (1978). The falling levels of the Caspian and Aral Seas, *Geog. Journal*, **144**, 62–80.

Homewood, K. M. (1976). Ecology and behaviour of the Tana River Mangabey (*Cercocebus galeritus galeritus*), unpublished Ph.D. Thesis, University of London.

Hotes, F. L. and Pearson, E. A. (1977). Effects of irrigation on water quality, in *Arid Land Irrigation in Developing Countries: Environmental Problems and Effects* (Ed. E. B. Worthington), pp. 127–58, Pergamon Press, Oxford.

Howarth, S. E. and Pant, M. P. (1987). Community managed irrigation in Eastern Nepal, *Irrigation and Drainage Systems*, **1**, 219–30.

Hughes, F. M. R. (1984). A comment on the impact of development schemes on the floodplain forests of the Tana River of Kenya, *The Geographical Journal*, **150**, 230–44.

Hughes, F. M. R. (1987). Conflicting uses for forest resources in the Lower Tana River basin of Kenya, in *Conservation in Africa: People, Policies, Practice* (Eds. D. M. Anderson and R. H. Grove), pp. 211–28, Cambridge University Press, Cambridge.

Hughes, F. M. R. (1988). The ecology of African floodplain forests in semi-arid zones: a review, *Journal of Biogeography*, **15**, 127–40.

Hughes, F. M. R. (1989). The influence of flooding regimes on forest distribution and composition in the Tana River floodplain, Kenya, *Journal of Applied Ecology* (in press).

Hulme, M. (1987). Secular changes in wet-season structure in central Sudan, *J. Arid Environments*, **13**, 31–46.

Ibrahim, F. N. (1982). The ecological problems of irrigated cultivation in Egypt, in *Problems of the Management of Irrigated Land in Areas of Traditional and Modern Cultivation* (Ed. H. G. Mensching), pp. 61–9, International Geographical Union, Hamburg.

ILACO (1973). *Tana River Feasibility Studies—the Bura Area*, International Land Development Consultants, Arnhem, The Netherlands.

ILACO (1975). *Bura Irrigation Scheme Feasibility Study, Final Report*, Report by International Land Development Consultants for FAO, Rome.

Jackson, I. J. (1977). *Climate. Water and Agriculture in the Tropics*, Longman, London.

Jayaraman, T. K. (1982). Operation and maintenance in surface irrigation projects in India, *Water Supply and Management*, **6**, 405–15.

Johansson, S. (1988). Forestry research in the Bura Irrigation Settlement project, Tana River District, Kenya, Bura Forestry Research Project Working Paper 37, University of Helsinki, Department of Silviculture and Kenyas Forestry Research Institute.

Jones, S. (1974). *Men of Influence in Nuristan*, Seminar Press, London.

Kassas, M. (1973). Impact of river control schemes on the shoreline of the Nile Delta, in *The Careless Technology: Ecology and International Development* (Eds. M. T. Farvar and J. P. Milton), pp. 179–88, Stacey, London.

Kenna, J. and Gillet, B. (1985). *Solar Water Pumping: A Handbook*, Intermediate Technology Publications.

Khogali, M. M. (1982). The problem of siltation in Khasm El Girba Reservoir: its implications and suggested solutions, in *Problems of the Management of Irrigated Land in Areas of Traditional and Modern Cultivation* (Ed. H. G. Mensching), pp. 96–106, International Geographical Union, Hamburg.

Kinawy, I. Z. (1977). The efficiency of water use in irrigation in Egypt, in *Arid Land Irrigation in Developing Countries: Environmental Problems and Effects* (Ed. E. B. Worthington), pp. 371–81, Pergamon Press, Oxford.

Kolarkar, A. S., Murthy, K. N. K. and Singh, N. (1983a). 'Khadin'—a method of harvesting water for agriculture in the Thar desert, *J. Arid Environments*, **6**, 59–66.

Kolankar, A. S., Singh, Y. V. and Lahiri, A. N. (1983b). Use of discarded plastic infusion sets from hospitals in irrigation on small farms in arid regions, *J. Arid Environments*, **6**, 385–90.

Kolawole, A. (1987). Environmental change and the South Chad Irrigation Project (Nigeria), *J. Arid Environments*, **13**, 169–76.

Kolawole, A. (1988). Cultivation on the floor of Lake Chad: a response to environmental hazards in eastern Borno, Nigeria, *Geog. Journal*, **154**, 243–50.

Kovda, V. A. (1977). Arid land irrigation and soil fertility: problems of salinity, alkalinity, compaction, in *Arid Land Irrigation in Developing Countries: Environmental Problems and Effects* (Ed. E. B. Worthington), pp. 211–35, Pergamon Press, Oxford.

Kowal, J. M. and Kassem, A. H. (1978). *Agricultural Ecology of Savanna: A Study of West Africa*, Clarendon Press, Oxford.

Law, J. P. Jr and Hornsby, A. G. (1982). The Colorado River salinity problem, *Water Supply and Management*, **6**, 87–100.

Ledec, G. (1987). Effects of Kenya's Bura Irrigation Settlement Project on biological diversity and other conservation concerns, *Conservation Biology*, **1**, 247–58.

Linares, O. F. (1981). From tidal swamp to inland valley: on the social organisation of rice production among the Diola of Senegal, *Africa*, **51**, 557–95.

Lowe-McConnell, R. H. (1975). *Fish Communities of Tropical Freshwaters*, Longman, London.

Ludwig, J. A. (1987). Primary productivity in arid lands: myths and realities, *J. Arid Environments*, **13**, 1–7.

McCleary, J. A. (1974). The biology of desert plants, in *Desert Biology* (Ed. G. W. Brown), 2 vols, pp. 141–95, Academic Press.

MacDonald, Sir M. (1977). Bura Project planning report, Sir M. MacDonald and Partners, Cambridge.

Marsh, C. M. (1976). *A Management Plan for the Tana River Game Reserve*, New York Zoological Society and University of Bristol, Bristol.

Martin, E. D. and Yoder, R. (1987). Institutions for irrigation management in farmer-managed systems: experience from the hills of Nepal, IIMI Research Paper 5, IIMI, Sri Lanka.

Masao, F. T. (1974). The irrigation system in Uchagga: an ethno-historical approach, *Tanganika Notes and Records*, **75**, 1–8.

Mensching, H. G. (Ed.) (1982). *Problems of the Management of Irrigated Land in Areas of Traditional and Modern Cultivation*. International Geographical Union, Hamburg.

Mitchell, D. S. (1977). Water weed problems in irrigation systems, in *Arid Land Irrigation in Developing Countries: Environmental Problems and Effects* (Ed. E. B. Worthington), pp. 317–28, Pergamon Press, Oxford.

Moris, J. R. and Thom, D. J. (1987). African irrigation overview: Main Report. Water Management Synthesis II, Report 37, Utah.

Moris, J., Thom, D. J. and Norman, R. (1984). *Prospects for Small-scale Irrigation*

Development in the Sahel, Water Management Synthesis II, Project Report 26, USAID, Washington.

Murray, S. P., Coleman, J. M., Roberts, H. H. and Salama, M. (1981). Accelerated currents and sediment transport off the Damietta Nile promontory, *Nature*, **293**, 51–3.

Nabhan, G. P. (1986). Papago Indian desert agriculture and water control in the Sonoran Desert 1697–1934, *Applied Geography*, **6**, 61–76.

O'Hea, A. (1983). Performance index for man-powered pumps, *Appropriate Technology*, **9**, 10–11.

Palmer-Jones, R. W. (1984). Mismanaging the peasants: some origins of low productivity on irrigation schemes in northern Nigeria, in *Irrigation in Tropical Africa: Problems and Problem-Solving* (Eds. W. M. Adams and A. T. Grove), pp. 98–108, African Studies Centre Monograph 3, Cambridge.

Palmer-Jones, R. W. (1987). Irrigation and the politics of agricultural development on Nigeria, in *State, Oil and Agriculture in Nigeria* (Ed. M. J. Watts), pp, 138–67, Institute of International Studies, Berkeley.

Park, C. C. (1981). Man, river systems and environmental impacts, *Progress in Physical Geography*, **5**, 1–31.

PEEM (1987). Effects of agricultural development and change in agricultural practices on the transmission of vector-borne disease, in *Report of the 7th Meeting of the Joint WHO/FAO/UNEP Panel of Experts on Environmental Management for Vector Control (PEEM)*, Rome, pp. 8–51, PEEM Secretariat.

Pels, S. and Stannard, M. E. (1977). Environmental changes due to irrigation development in semi-arid parts of New South Wales, Australia, in *Arid Land Irrigation in Developing Countries: Environmental Problems and Effects* (Ed. E. B. Worthington), pp. 171–83, Pergamon Press, Oxford.

Petts, G. E. (1984). *Impounded Rivers: Perspectives for Ecological Management*, Wiley, Chichester.

Pike, A. G. (1965). Kilimanjaro and the furrow system, *Tanganika Notes and Records*, **64**, 95–6.

Richards, P. (1985). *Indigenous Agricultural Revolution: The Ecology of Food Production in West Africa*, Hutchinson, London.

Richards, P. (1986). *Coping with Hunger*, Allen and Unwin, London.

Roger, R. A. (1984). The Dallou irrigation system, *Prospects for small scale irrigation development in the Sahel* (Ed. J. Moris), (USAID Water Management Synthesis II Project, Utah, 1984), pp. 155–57.

Roggeri, H. (1985). *African Dams: Impacts on the Environment*, Environment Liaison Centre, Nairobi.

Ron, Z. Y. D. (1985). Development and management of irrigation systems in mountain regions of the Holy Land, *Trans. Inst. Br. Geogr.*, **10**, 149–69.

Saad, M. A. H. (1982). Distribution of nutrient salts in the lower reaches of the Tigris and Euphrates, Iraq, *Water Supply and Management*, **6**, 443–53.

Schilfgaarde, J. van (1982). The Welton–Mohawk Dilemma, *Water Supply and Management*, **6**, 115–27.

Shalash, S. (1983). Degradation of the River Nile, Parts 1 and 2, *Water Power and Dam Construction*, **35**(7), 37–43 and **35**(8), 56–8.

Singh, D. V., Pal, B., Pal, K. and Kishore, R. (1981). Quality of underground irrigation waters in semi-arid tract of district Agra. III—tehsils Bah and Fatehabad, *Annals of Arid Zone*, **20**, 48–52.

Skogerboe, G. V., Walker, W. R. and Evans, R. G. (1982). Salinity control measures for Grand Valley, *Water Supply and Management*, **6**, 129–67.

Sutton, J. E. G. (1978). Engaruka and its waters, *Azania*, **13**, 37–70.

Sutton, J. E. G. (1984). Irrigation and soil conservation in African agricultural history: with a reconsideration of the Inyanga terracing (Zimbabwe) and Engaruka irrigation works (Tanzania), *J. Af. Hist.*, **25**, 25–41.

Thomas, G. A. and Jakeman, A. J. (1985). Management of salinity in the River Murray Basin, *Land Use Policy*, **2**, 87–102.

Underhill, H. W. (1984), *Small Scale Irrigation in Africa in the Context of Rural Development*, FAO Land and Water Development Division, Rome.

Urroz, E. (1977). Reutilisation of residual water and its effects on agriculture, in *Arid Land Irrigation in Developing Countries: Environmental Problems and Effects* (Ed. E. B. Worthington), pp. 361–70, Pergamon Press, Oxford.

Vainio-Matilla, A. (1987). *Domestic Fuel Economy*, Bura Fuelwood Project, University of Helsinki Institute of Development Studies Report 13, Helsinki.

Vogel, H. (1988). Impoundment-type bench terracing with underground conduits in Jibil Hiraz, Yemen Arab Republic, *Trans. Inst. Br. Geogr. N.S.*, **13**, 29–38.

Wade, R. (1982). The system of administrative and political corruption: canal irrigation in South India, *J. Dev. Studies*, **18**, 287–328.

Wallace, T. (1979). Rural development through irrigation: studies in a town on the Kano River Project, Report 3, Centre for Social and Economic Research, Zaria, Nigeria.

Wallace, T. (1980). Agricultural projects and land in northern Nigeria, *Review of African Political Economy*, **17**, 59–70.

Wallace, T. (1981). The Kano River Project: the impact of an irrigation scheme on productivity and welfare, in *Rural Development in Tropical Africa* (Eds. J. Heyer, P. Roberts and G. Williams), pp. 281–305, Macmillan, London.

Warren, A. and Agnew, C. (1988). *An Assessment of Desertification and Land Degradation in Arid and Semi-arid Areas*, IIED, London.

Warren, A. and Maizels, J. K. (1977). Ecological change and desertification, in *Desertification: Its Causes and Consequences* (Ed. Secretariat of the UN Conference), pp. 169–261, Pergamon Press, Oxford.

Watt, S. B. (1981). Peripheral problems in the Senegal Valley, in *River Basin Planning in Theory and Practice* (Eds. S. K. Saha and C. J. Barrow), Wiley, Chichester.

Welcomme, R. L. (1979). *Fisheries Ecology of Floodplain Rivers*, Longman, London.

Wellings, F. M. (1982). Viruses in groundwater, *Environment International*, **7**, 9–14.

Wilkinson, J. C. (1974). The organisation of the Falaj irrigation system in Oman. Oxford School of Geography Research Papers 10.

World Bank (1977). *Bura Irrigation Settlement Project Appraisal Report*, World Bank, Washington.

World Bank (1981). *Accelerated Development in Sub-Saharan Africa*, World Bank, Washington.

World Bank (1985). *Bura Irrigation Settlement Project Mid-term Evaluation Report 1984*, Main Report, unpublished ms.

Worthington, E. B. (Ed.) (1977). *Arid Land Irrigation in Developing Countries: Environmental Problems and Effects*, Pergamon Press, Oxford.

Yair, A. (1983). Hillslope hydrology water harvesting and areal distribution of some ancient agricultural systems in the northern Negev Desert, *J. Arid Environments*, **6**, 283–301.

CHAPTER 6

Water Erosion and Conservation: An Assessment of the Water Erosion Problem and the Techniques Available for Soil Conservation

R. LAL

INTRODUCTION

Soil erosion by water is a work process involving two phases: detachment of soil particles and their transport. The energy required for these two work functions is supplied through raindrop impact, overland flow, and the interaction between them. Soil detachment involves removal of transportable fragments of material from a soil mass by raindrop impact or shearing force of overland flow. Transport or entrainment of detached primary or secondary particles occurs through splash and overland flow.

Soil Detachment

Detachment of particles by raindrop impact from a dry soil clod occurs through three different stages (Yariv, 1976): dry soil, soil–water mixture or fluidized soil, and soil-cum-overland flow. The collision between dry soil particles and raindrops is essentially a collision of two elastic bodies. As soil strength decreases due to progressive increase in soil-water potential, primary particles are easily splashed from the fluidized soil. The maximum splash usually occurs during the second stage of wetness. Soil detachment depends on many factors as shown in Figure 6.1.

Forces promoting soil detachment are those related to the rainfall and overland flow. Rainfall characteristics of importance in soil detachment are

Techniques for Desert Reclamation
Edited by A. S. Goudie
© 1990 John Wiley & Sons Ltd.

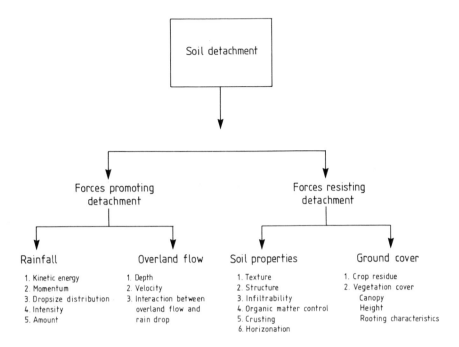

Figure 6.1   Factors affecting soil detachment

drop size distribution, and the angle and velocity of its impact. The latter
depends on whether the rain is wind-driven or not. Kinetic energy of the rain
is the principal factor responsible for soil detachment, although momentum is
also considered important. There are two separate forms of kinetic energy
(Kinnell, 1981):

(a) the rate of expenditure of the rainfall kinetic energy which has the units
     of energy per unit area per unit time (E_{RR}) and
(b) the amount of rainfall kinetic energy expended per unit quantity of rain
     that has the units of energy per unit area per unit depth (E_{RA}).

Both types of energy are interrelated (that is $E_{RA} = c\,E_{RR}^{-1}$). The amount
of rainfall kinetic energy per unit area per unit time is related to rainfall
intensity. A commonly used empirical relation is

$$E_{RA} = a + b \log I \tag{1}$$

where I is rainfall intensity and a and b are empirical constants.

The depth of overland flow interacts with the drop diameter in causing soil
detachment. The turbulence caused in the water by impacting raindrops
increases the detaching capacity (Palmer, 1963; Mutchler and Larson, 1971;
Walker *et al.*, 1977).

Soil Transport

Factors affecting soil transport are listed in Figure 6.2. The depth of concentrated flow in rills may be 50 times that of overland flow, and flow velocity may be as much as 10 times greater (Young and Wiersma, 1973). The energy of flow depends on its velocity. There is a critical velocity, corresponding to each soil type below which significant movement of particles does not begin.

Slope steepness affects particle transport because of its impact on runoff velocity. Soil erosion is related to soil steepness by a logarithmic or power function with the slope exponent varying between soils (Meyer and Monke, 1965; Kirkby, 1971; Lal, 1976; Morgan, 1978). The effect of slope length on erosion is governed by slope gradient and shape. With a regular slope, soil erosion is generally believed to increase proportionately to some power of slope length (Kramer and Meyer, 1969; Laflen *et al.*, 1978; Foster *et al.*, 1977; Mutchler and Greer, 1980; Lal, 1988a). Soil erosion is usually more on convex than on regular or concave slopes (Lal, 1976).

Soil properties that influence sediment transport include particle size, soil strength, and water transmission characteristics. Apparently clay- and silt-

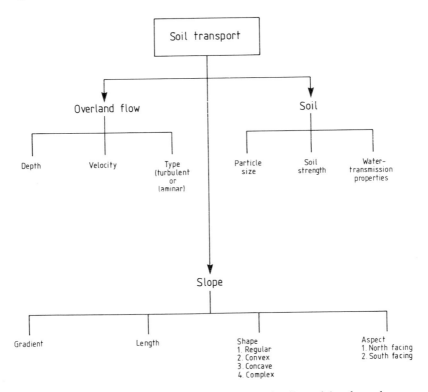

Figure 6.2   Factors affecting displacement of soil particles downslope

sized particles are more easily transported than sand-sized particles. Further-more, a soil that can resist the shearing force of running water is less easily transported or displaced. Soils that have a high water-intake rate have less overland flow, and are relatively less susceptible to erosion than those with a restricted or slow intake rate.

ENVIRONMENTS, HYDROLOGY AND SOILS OF ARID REGIONS

The arid and semi-arid regions of the world cover about 4885 million hectares or 33% of the total land area (Table 6.1). Accelerated erosion, by wind and water, can be a severe problem in these regions. In fact, erosion-induced soil degradation and desertification are widespread ecological problems in arid regions. Desertification, or the process leading to expansion of the arid climate to the adjoining ecological regions, is a severe problem in Africa and the Middle East (Table 6.2). The spread of desert-like conditions is caused by accelerated erosion, denudation of vegetation cover, and reduction in effec-tiveness of the precipitation received. The severity of soil erosion by water in arid climates is related to many factors, some of which are briefly outlined below.

Table 6.1   The world's arid lands

Category	Area $(10^8$ ha$)$	Percentage of world land area
Semi-arid	21.24	14.3
Arid	21.80	14.7
Extremely arid	5.81	3.9
Total	48.85	32.9

Modified from Matlock (1981).

Table 6.2   Desertification in the developing countries

Region	Area $(10^8$ ha$)$				
	Slight	Moderate	Severe	Very severe	Total
Africa	5.154	7.948	1.468	0.038	14.608
Asia	0.068	0.216	1.011	0	1.295
Latin America	0.092	0.398	1.789	0.015	2.294
Middle East	0.605	0.357	9.516	0	10.479

Modified from Matlock (1981).

Climate

Arid regions are characterized by rainfall of less than 400 mm per year. The rainfall is concentrated during 2 or 3 months, and its effectiveness is extremely low. Many climatic indices have been developed to characterize arid climates in terms of their rainfall effectiveness (Table 6.3). According to Budyko's Index of Dryness, the value of this index for arid regions is usually greater than 5 and that for semi-arid regions lies between 3 and 5. Annual precipitation is an important component in all indices. However, an important climatic index relevant to soil erosion by water is the one that addresses the issue of rainfall effectiveness, rainfall intensity, and the rate and amount of overland flow. Although none of the indices listed in Table 6.3 involves runoff, indices proposed by De Martonne, Meyer, Thornthwaite, and Capot-Rey address the problem of rainfall effectiveness. None of the indices is concerned with the rainfall intensity or its energy load.

These climatic indices are indicative of the annual water balance, especially from the point of view of crop growth and vegetation cover. Rainfall parameters, such as amounts and intensities, are also important in erosion by water. Torrential rains, with short-term intensities exceeding 100 mm/h, are not uncommon for these regions. Furthermore, rainfall is concentrated during 2 or 3 months and flash floods from denuded hills cause severe erosion. Rainfall in arid regions is highly localized, and is limited both in space and duration.

Runoff

Water runoff in arid regions is characterized by flash floods of high velocity but short duration. The rapid flow is attributed to low soil infiltrability, a high proportion of rain converted to overland flow, and scanty or no vegetation cover to retard the flow velocity. Most runoff events are localized, with a limited quantity flowing to ocean (Table 6.4). Concentrated or channelized flow in arid regions is intermittent, erratic, and irregular because streams do not feed on spring water.

Surface runoff created by isolated and localized storms has high velocity. In the absence of any protective vegetation cover, the soil is subjected to the high shearing force of such an overland flow. Consequently, runoff in arid regions is characterized by high turbidity values. The specific turbidity, suspended load per cubic metre, increases from a minimum in the cold climate, through the temperate, the equatorial, and the Mediterranean, until maximum values are attained in the semi-arid climate. Nir (1974) reported specific turbidity values of 20 g/m^3 for the Siberian Yenisei, 20 kg/m^3 in the Chaco River in northern Argentina, 78 kg/m^3 in the Little Colorado, and 144 kg/m^3 in the Rio Grande del Norte in Brazil.

Table 6.3   Climatic indices to characterize arid climates

Author	Index
1. Köppen	Rainfall amount versus mean annual temperature
2. Lang	Rain Factor Index $= \dfrac{P}{T}$
3. De Martonne	The Index of Aridity $(I) = \dfrac{P}{T + 10}$
4. De Martonne	$I' = \dfrac{P/(T + 10) + 12p/(t + 10)}{2}$
5. Meyer	Precipitation–saturation deficit ratio $= \dfrac{P}{SD}$
6. Thorthwaite	Precipitation Effectiveness Index: $= P'/E$ Annual P'/E is obtained by determining the sum of the P'/E ratio for each month of the year and multiplying by 10.
7. Modified Thorthwaite Index	$\dfrac{P'}{E} = \displaystyle\sum_{i=1}^{12} 115 \left(\dfrac{P'}{T' - 10}\right)^{9/10}$
8. Capot-Rey	Rainfall Efficiency Index $= \dfrac{100P/E + 12p'/e'}{2}$
9. Budyko (1974)	Index of Dryness $= \dfrac{R \ (l/yr)}{L \times P''}$

where P = mean annual precipitation (mm) E = evaporation (in)
        T = mean annual temperature (C°) T' = mean monthly temperature (F°)
        p = total annual rainfall of the driest p' = rainfall of the wettest month (mm)
            month (mm) e' = evaporation of the wettest month
        t = average temperature of the driest   (mm)
            month (C°) L = latent heat of vaporization
       SD = saturation deficit (mmHg)   (600 cal/g)
        P' = precipitation (in) P' = mean annual precipitation (cm)

Modified from Lal (1987) and Nir (1974).

Soils

Predominant soils of arid region are Entisols, Aridisols, Mollisols, Alfisols and Vertisols (Table 6.5).

   Entisols are mineral soils of recent origin showing little or no evidence of horizonation. These soils are young and are derived from recent deposits.

Table 6.4   Area of arid lands based on surface runoff

Runoff		Area (10^8 ha)
No surface runoff		27.991
Land surface runoff not reaching oceans		13.847
	Total	41.838

Modified from Dregne (1976).

Table 6.5   Predominant soils of the arid regions

Soil order	Area (10^8 ha)	Percentage of world land area
Entisols	19.149	13.1
Aridisols	16.570	11.3
Mollisols	5.475	3.7
Alfisols	3.070	2.1
Vertisols	1.885	1.3
Total	46.149	31.5

Modified from Dregne (1976).

Most common Entisols in arid regions are Psamments. They have sandy texture and are easily eroded.

Aridisols are characterized by low organic matter content and prevalence of desert pavement of pebbles, gravels, or stones. The pavement is caused by the past erosion of fine soil particles, leaving the coarse fragments on the soil surface.

Mollisols occur in depressions and valleys. These soils have a dark-coloured surface horizon and have a relatively high organic matter content. These soils are less erodible.

Alfisols have a well-defined argillic or clayey horizon, and have a high base saturation percentage. Some Alfisols may also have a fragipan, duripan, natric horizon, or hardened plinthite. Alfisols are usually well drained and have a reddish colour. Alfisols have a high susceptibility to erosion by water.

Vertisols are dark, clayey soils with a clay content of 30% or more. These soils have a high swell/shrink capacity, low infiltration, and high runoff. When dry these soils develop large and deep cracks. The cracks usually open and close more than once in most years. These soils occur in monsoonal climates and are highly susceptible to erosion.

Soil properties responsible for high susceptibility of arid region soils to erosion by water are particle size distribution, structural stability, and crusting.

Particle size distributions: Arid soil containing predominantly silt and medium sand are highly susceptible to erosion by water. Particles most easily detached by raindrop impact and running water are in the $250–1000\,\mu$m range. Unaggregated coarse particles are easily detached and washed away.

Soil organic matter content   The organic matter content of soils of the arid regions is generally low. Consequently, soil particles are not cemented together by organic matter. Only Mollisols have a sufficient level of soil organic matter content to favourably influence soil structure.

Structural stability . The erosion rate depends on the resistance of the aggregates to dispersion and on the size of the water-stable aggregates for resistance to entrainment by splashing and runoff. Being low in organic matter content, soils of arid regions are generally less aggregated. In addition to a low organic matter content, the soil's ability to resist detachment also depends on the amount and nature of the clay and on the nature of cations on the exchange complex. Soils with a natric horizon are easily dispersed due to the predominance of Na^+ on the exchange complex.

Crusting   In most arid and semi-arid areas the overland flow is generated by the presence of surface crust. The crust or surface seal is usually 1 to 2 mm thick and has an extremely low permeability to water. It is formed by: (a) the slaking of aggregates due to raindrop impact, (b) the dispersion of surface soil, and (c) rapid drying due to high evaporative demand. Sizeable runoff is generated by high-intensity, torrential rains falling on crusted soils denuded of any effective vegetation cover.

Hydrophobicity   Many soils of the arid regions exhibit 'hydrophobic' or water-repellent characteristics. The hydrophobic characteristic is attributed to the large contact angle between the solid/liquid interface. The water repellence is imparted by a thin coating of specific inorganic or organic chemicals. Burning of shrubs and biomass is known to increase water repellence. Hydrophobic soils generate high runoff and are prone to severe erosion.

Infiltration rate   Infiltration, the volume flux of water flowing into the profile per unit of soil surface area, is generally low for soils and semi-arid regions. Low aggregation and poor stability of aggregates, water repellence, crusting, and the presence of desert pavement are some of the factors responsible for a low infiltration rate of soils in the dry regions.

Land Use and Farming Systems

Shifting cultivation and other extensive types of agriculture are the predominant types of farming systems of arid regions. Traditional farming systems range in diversity from nomadic herding to semi-permanent rainfed agriculture or permanent irrigated agriculture. Mixed farming is common. Cropping intensity in rainfed agriculture is low and the land-use factor (L)* is generally between 5 and 10. Predominant types of farming systems are shown in Figure 6.3. The erosion hazard is generally more in dryland than in irrigated farming, in extensive than in intensive cropping, and in monoculture than in mixed cropping systems.

ASSESSMENT OF SOIL EROSION

The data based on field measurements of runoff and erosion from arid regions are scanty. In addition, information on soil, climate, hydrology, landscape, and physiography is also limited. It is therefore difficult to make reliable estimates of the soil erosion hazard. Soil erosion can be assessed by visual observations, field measurements, and through modelling. Methods of assessment of soil erosion are described in detail by Lal (1988b).

Visual Assessment

Visual assessment of soil erosion is rather subjective. Information of this nature, although useful, is only of relative importance. Assessment by visual observations is generally qualitative and based on reconnaissance surveys. A

---

*The land-use factor (L) is defined as the ratio of cropping period C plus fallow period F to cropping period $C[L = (C + F)/C]$.

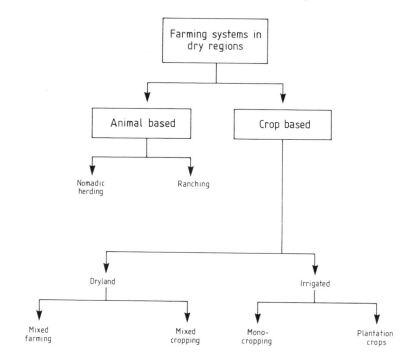

Figure 6.3   A generalized classification of farming systems

rating system is used to assess the degree of soil erosion. One such system developed by USDA (1951) is shown in Table 6.6.

Field Measurements

Field measurements may involve runoff plots, small or large watersheds, and river gauging.

Field plot techniques   Runoff plots are established on appropriate soil and slopes to monitor the magnitude of soil erosion in relation to soil properties, land form, and land use. Such studies, often long term and capital intensive, provide the most reliable information. Different types of plots and the equipment needed to monitor runoff and erosion are described by Mutchler *et al.* (1988) and Meyer (1988), and are briefly described in Table 6.7.

It is extremely important to standardize the methods used. Assessment of soil erosion must be based on techniques that produce reliable, precise, and

Table 6.6   A rating system developed by USDA to assess the degree of soil erosion (USDA, 1951)

Class	Description
1	No apparent or slight erosion
2	Moderate erosion: moderate loss of topsoil generally and/or some dissection by runoff channels or gullies
3	Severe erosion: severe loss of topsoil generally and/or marked dissection by runoff channels or gullies
4	Very severe erosion: complete truncation of the soil profile and exposure of the subsoil (B horizon) and/or deep and intricate dissection by runoff channels or gullies

reproducible results. The data should be collected for at least 3 years to establish long-term trends. An example of such a field study is that established at the Arab Centre for the Studies of Arid Zone and Dry Lands (ACSAD) in Syria. Thirty-six runoff plots were established in marly sedimentary hills of the coastal zone of Syria about 25 km south-east of Al-Latakiya. Such long-term, well-designed, and properly equipped field studies are needed to obtain reliable data.

Establishing sediment load in streams and rivers   Measurement of sediment transport in rivers is an alternative technique to assess the average rate of soil erosion over the entire watershed. Techniques for measurement of sediment yield from large watersheds are described by Walling (1988). Walling has also outlined major problems with any attempt to use sediment yield data to assess on-site rates of erosion and soil loss within a watershed. An important problem is the lack of information on the sediment delivery ratio (SDR). SDR is the ratio of sediment delivered at the watershed to the gross erosion within the watershed. Considerable uncertainty surrounds the information available and the methods used for calculating the SDR. Methods used in assessment of the sediment loads are also unstandardized.

Reservoir siltation   The siltation rates of reservoirs are also used as a method to assess the average erosion over the watershed (Le Houérou, 1970; Monjauze, 1960).

Satellite imagery and cartographic analysis   Remote sensing techniques are increasingly being used to assess soil erosion. Hamza and El-Amami (1977)

Table 6.7 Assessment of soil erosion by field experimentation

Plot	Size	Objectives	Rainfall	Runoff collection system
Microplots	<25 m^2	To compare relative erosion	Natural/simulated	Collect the entire amount by storage tank
Field plots	25–500 m^2	To evaluate soil and crop management systems	Natural/simulated	Multidivisor tanks, flume, water stage recorder, sampling device
Hill slope	50–100 m long	To assess erosion on long slopes	Natural	Gerlach traps
Agricultural/ farmland	1–2 hectares	To assess the effects of cropping/farming systems	Natural	Weirs, V-notch, H-flume, water stage recorder, sampling device
Large watersheds	5–1000 hectares	To evaluate the effects of land use	Natural	Water stage recorder, river gauging, sediment sampling

prepared maps of soil and erosion/deposition for Tunisia using remote sensing techniques.

Radioisotopes and other tracers   Radiotracer techniques have been used to identify the source of sediment over the watershed (Ritchie *et al.*, 1974, 1975). The environment tracer caesium-137 (^{137}Cs) has a half-life of 30 years and can be used to estimate erosion or deposition rates over small delineated watersheds.

Measuring changes in soil level   Remnants of the original soil surface, pedestals formed by stone covers, and exposed tree roots are used to estimate the depth of soil eroded. Extreme caution is needed to assess erosion from temporal changes in soil level.

Predicting Soil Erosion   Soil erosion can be predicted knowing the factors of climate, soil, topography, and land use.

The universal soil loss equation   This parametric equation was developed in the late 1950s, and has been the most widely used tool for predicting soil erosion (Wischmeier and Smith, 1958). The equation is

$$A = RKLSCP$$

where A is the average annual soil loss, R is the erosivity of rainfall, K is soil erodibility, L is slope length, C is cropping management, and P is conservation practice. The USLE is based on more than 25 years of research covering 10 000 plot-years of data from a natural runoff plot, and the equivalent of 1000 plot-years of data from field plots using rainfall simulators. The ecological limits of the data base of the USLE are, however, narrow because the experiments were limited to the midwestern USA.

The USLE estimates the average soil loss over an extended period, i.e. average annual soil loss. The equation does not predict storm-by-storm soil loss. This equation does not apply under the following conditions:

(a) in geographical regions where basic information on various factors (R, K, C, and P) is not available or factor values cannot be accurately derived from existing data,
(b) for predicting soil erosion from complex watersheds by taking average slope length and making other adjustments,

(c) for estimating soil erosion from slopes steeper than 20%, for regions with high torrential rains and for highlands, and
(d) for estimating soil erosion from specific rain events.

Various modifications have been proposed. Modifications in the method for assessment of factor R have been proposed by Hudson (1965), Elwell and Stocking (1975), Lal (1976), FAO (1979), and Zanchi and Torri (1980). Williams (1975) proposed modifications in USLE to estimate sediment yield from individual runoff events from a watershed by modifying the factor R. The modified equation is called MUSLE. The SCS has also undertaken revision of the equation and it is called the revised USLE or RUSLE. Estimates of the soil erodibility factor K from the nomogram proposed by Wischmeier *et al.* (1969) are also not reliable for many soils of the tropics (Vanelslande *et al.*, 1987).

The upland erosion model Foster and Meyer (1975) and Foster (1982) proposed the 'inter-rill' and 'rill' erosion model. The model computes sediment contributions for inter-rill and rill regions separately and then adds these together to estimate the upland erosion. It assumes a uniform down-slope flow, and that flow and sediment are uniformly distributed across the slope.

The Griffith University model Rose (1988) proposed another process-oriented model. This model assumes that land slopes and soil strength relationships are such that land slides, mud flows, and gully erosion do not occur, and that upland erosion is mainly associated with the shallow overland flow and concentrated rill flow.

Other predictive models Knisell (1980) proposed another model to assess 'Chemical Runoff and Erosion from Agricultural Management Systems' (CREAMS). It is designed to evaluate non-point source pollution from field-sized areas. The Soil Conservation Service (SCS) of the USDA is now developing another model called the 'Water Erosion Prediction Project' (WEPP).

Emphasis should be given to developing those models which build on the physical laws and nature of the system. These models should have well-defined equations for computing detachment by raindrop impact, detachment by overland flow, transport by splash and overland flow, and deposition by overland flow. The successful model should enable computations of these components for each identifiable segment of the watershed. WEPP is designed to meet these objectives.

FACTORS RESPONSIBLE FOR ACCELERATED SOIL EROSION IN ARID REGIONS

Arid lands are fragile ecosystems. The delicate ecological balance is preserved as long as human intervention does not disturb the harmonious coexistence of plants, animals, birds, and insects. However, demand by increasing human and animal population has stressed this fragile ecosystem. Accelerated soil erosion by water is a consequence of this stress. Some factors responsible for severe erosion are briefly described below.

Intensive Farming

Removal of protective vegetation cover, by uncontrolled grazing or for an intensive arable land use, is a major factor responsible for accelerated soil erosion. Changing patterns of food consumption have caused alterations in land use resulting in a severe erosion hazard in steeplands of arid and semi-arid climates. When the population was low, the nomadic grazing of the high plateaux was complementary to the more settled pastoral farming of the lowlands. The increase in population pressure, however, has led to the expansion of arable farming and overgrazing of the plateaux lands. The nomadic habits have been abandoned which has led to severe grazing pressure on steeplands (Boukhobza, 1982). Accelerated soil erosion in arid lands has been known to cause severe ecological problems throughout human history. A reason for the fall of the Empire of Mesopotamia is cited to be the siltation of irrigation canals by sediments transported from the cutover forest and overgrazed hill lands. In recent times, the decline in terrace cultivation as a result of poor maintenance has caused irreversible erosion damage to the land in Haraz Province of the Yemen Arab Republic (Alkamper *et al.*, 1979). In Tunisia, Floret and Le Floch (1977) observed that overgrazing caused 88.5 t/ha of soil loss in about 5 months. Erosion rates, however, declined drastically by restoring vegetation cover through controlled grazing. Removal of protective vegetation cover for fuelwood or forage has caused severe erosion in the lower watershed of the Wad El-Hadjeh region of Tunisia (Bonvallot and Hamza, 1977).

Soil Erodibility, Parent Material, and Terrain

Detachability and transportability of soil by erosion agents depend on the physical, chemical, and biological properties of a soil. Soil erodibility, or susceptibility of soil to erosive action of rain and overland flow, depends on soil properties. Because of differences in soil properties, some soils are more susceptible than others. In Turkey, Balci (1973) observed that organic matter, gravel, and clay contents were inversely related with the dispersion ratio. In contrast, electrical conductivity and silt content were positively correlated with the dispersion ratio. In general, the higher the dispersion ratio the more the erodibility. The dispersion ratio, the amount of easily dispersible clay in relation to the total clay content, is influenced by the soil organic matter content and land use. Soils with a dispersion ratio exceeding 15% are usually susceptible to erosion by water. All other factors remaining the same, the dispersion ratio is in the order of row crop farming ⩾ intensively grazed pastures > well-grazed pasture > undisturbed scrub or desert vegetation.

Another relevant index of erodibility is the percent aggregation and the water stability of aggregates. For arid soils, $CaCO_3$ plays an important role in aggregation. In Egypt, Ghazi (1982) observed that the mean weight diameter based on both dry and wet sieving tests and aggregates exceeding 0.2–0.4 mm diameter were significantly correlated with hydraulic conductivity. Soils with a high saturated hydraulic conductivity have a lower runoff than those with a low hydraulic conductivity.

The parent material affects erodibility through its influence on soil properties. In Turkey, for example, Balci (1973) and Balci and Ozyuvaci (1974) observed that the erodibility of soils in the region of north-central Anatolia was in the order of Neogene silt > sandstone > andesite > conglomerate. For soils in the Kocaeli region, erodibility was in the order of soil derived from sandstone > sand and gravel > shale > limestone. Furthermore, soils developed on southern slopes are more susceptible than those developed on northern slopes. Soils derived from lime-rich marl have a low infiltration rate and are susceptible to erosion. That is why the annual suspended sediment yield in North African rivers is the highest for those flowing through marl than for those flowing through sandstone, limestone, or dolomite (Walling, 1988).

In addition to the nature of the parent material, the slope gradient and aspects also have a significant effect on sediment transport. In Tunisia, Chaabouni (1977) reported significant effects of slope gradient on sediment yield.

Seasonal Water Balance and Rainfall Variability

Water deficit influences erosion by water both directly and indirectly. Directly it influences soil water retention and transmission on the one hand and runoff rate and amount on the other. Indirectly, water deficit affects soil organic matter content and vegetation cover. In general, the drier the climate, the greater is the water erosion caused by torrential rains. High-velocity winds and low relative humidity, a characteristic feature of these regions, affect soil moisture content and crop growth. With the exception of the short rainy season, the relative humidity is usually less than 10%. These desiccating winds affect both soil properties and vegetation cover.

Precipitation in the arid regions is highly variable. In the Oglat Merteba region of Tunisia, for example, Floret *et al.* (1977) reported that while the mean average rainfall is 183 mm the range is from 39 to 460 mm. Furthermore, as much as 50 to 60% of the annual rainfall can be received in a 24-hour period, causing high runoff and severe flash floods. In Egypt, Labib (1981) and Renard (1982) reported severe erosion damage due to flash floods caused by high-intensity isolated rainstorms. Flash floods caused by exceptional rains are an important cause of severe erosion in arid regions. Gully erosion is drastically accentuated by flash floods.

Rainfall intensity is another important factor to be considered. Rainstorms in dry regions can have high intensities over a short period of time. Rainfall intensity is related to the duration and recurrence interval or return period of the storm. Finkel and Finkel (1986) reported the relationship of rainfall intensity (I) in millimetres per hour to the return period (T) in years and duration of the given intensity (t) in minutes for Jerusalem, Israel, as shown:

$$I = \frac{15T^{0.2}}{t^{0.51}} \tag{2}$$

Perpetual water deficit and highly eratic rainfall are not conducive to high biomass production (Goudie and Wilkinson, 1977). Denuded landscape, soils highly susceptible to erosion, and flash floods lead to severe erosion because even a small increase over the mean seasonal rainfall produces a disproportionate increase in runoff. Runoff from arable fields may be as much as 30 to 50% of the rainfall received (Agassi *et al.*, 1986).

Finkel and Finkel (1986) studied the occurrence of floods on different tributary catchments of the Arark Valley, Israel. They observed that floods on all catchments followed the Poisson distribution. This implies that each flood, on any stream, was entirely a random chance occurrence and independent of any other floods on that stream or any other stream in the area. The frequency of floods was high and ranged between 1.23 and 2.44 floods per year on each stream.

Tunnel Erosion and Soil Fauna

Piping or tunnel erosion is caused by rapid and concentrated sub-surface flow. Pipe flow is a universal phenomena in which the flow velocities are of the same order of magnitude as overland flow. Pipe flow is generally fed by cracks flow or flow through animal burrows. After tunnel erosion has occurred, the soil surface may collapse leading to the initiation of a gully. Some soils of the arid and semi-arid regions are highly susceptible to tunnel erosion. Tunnel erosion is particularly severe in soil with high contents of $CaCO_3$ and other water-soluble salts.

Activity of soil animals is known to play a major role in soil erosion in arid and semi-arid regions. Earthworms, termites, and rodents have a significant

Table 6.8   Sediment produced by the captivity of Isopods and Porcupines in the desert of northern Negev, Israel

Location	Annual rainfall (mm)	Available sediment production (kg/ha)	
		Isopods	Porcupines
1	310	413	302
2	110	45	35
3	100	53	33
4	93	119	87
5	85	21	66
6	65	0	29

Modified from Yair and Rutin (1981).

effect on soil turnover and its water-transmission properties. Although burrowing and digging activities of these animals have beneficial effects on soil properties, they also loosen the sediment for easy transport with the overland flow. The data by Yair and Rutin (1981) shown in Table 6.8 indicate that a large amount of sediment is produced by the activity of Isopods and Porcupines.

TECHNIQUES AVAILABLE FOR SOIL CONSERVATION

Soil conservation involves both erosion preventive and control measures. Erosion prevention is often more effective than control. Some principles of soil conservation are outlined in Figure 6.4. Erosion preventive measures

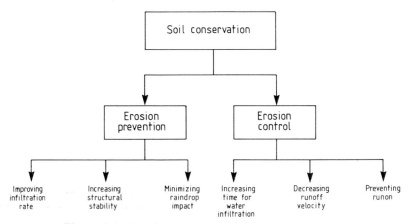

Figure 6.4   Erosion prevention versus control measures

Table 6.9 Technique to reduce soil splash and inter-rill/rill erosion

Principle	Technique	Cultural practices
1. Prevent raindrop impact	Maintain a cover close to the ground surface	• Afforestation • Crop residue management • Multistorey canopy • Live mulch
2. Increase infiltrability	Improve macroporosity	• Chiselling • Sub-surface tillage • Mulching
3. Reduce runoff velocity	Rough surface	• Rough seedbed • Mulching

involve cultural practices that minimize raindrop impact, increase or enhance structural stability of the soil, and improve the water intake rate or infiltrability. Techniques that control or reduce erosion involve management of surplus water or overland flow for its safe disposal at low velocities. Techniques with potential to reduce sheet wash and rill erosion are shown in Table 6.9 and those with applications towards runoff management are listed in Table 6.10.

Erosion Preventive Measures

Accelerated soil erosion is a symptom of land misuse. Erosion becomes a severe problem when land is used beyond its capacity. Erosion preventive measures are, therefore, those cultural techniques that involve a rational land use. The latter is based on the land capability assessment. Soil erosion is generally within the tolerable limits when land is used for whatever it is capable of and by those methods of soil and crop management that are ecologically compatible.

Land use Marginal lands, soils of steep gradient and shallow rooting depth, must not be used for grain crop production or for excessive and uncontrolled grazing. Such lands should be protected and kept under natural vegetation cover. Ungrazed woodlands, scrub vegetation, or grasslands protect the soil against raindrop impact and the shearing force of running water. An actively frowing vegetation cover also maintains a favourable rate of infiltration. Frequency of occurrence and severity of flash floods is drastically reduced from vegetated hill slopes. Although coercive measures are rarely effective,

Table 6.10 Possible techniques for decreasing runoff volume and peak runoff rate

Soil	Principle	Technique
Structurally unstable coarse-textured soils, in semi-arid regions	Prevent surface sealing and raindrop impact	Mulching, reduced tillage, cover crop
Soils with low activity clays and coarse texture, in semi-arid to arid regions	Increase surface detention	Rough cloddy surface by ploughing at the end of rainy season
Loamy to clayey-textured soils, easily compacted	Improve infiltration	Vertical mulching, chiselling
Good-structured soils, with well-aggregated, stable structure	Prolong draining time	Tied ridges, contoured ridge-and-furrow system
Vertisols; soils with expanding clay minerals in arid regions with short rainy season	Maintain soil surface at moisture potential above the hygroscopic coefficient and reduce heat of wetting	Mulching, soil inversion just prior to rain
Vertisols; soils with expanding clay minerals in semi-arid or sub-humid regions with long rainy season	Dispose of water safely and recycle for supplementary irrigation	Graded ridge-and-furrow system with grass waterways and storage tank, camber bed technique

the general public should be made aware of the dangers of misusing the marginal steeplands. These lands should be protected against bush fire, grazing, and wood cutting and developed only for recreational purposes.

Farming systems   The farming system is a resource management strategy designed to optimize the economic returns without jeopardizing the resource base. Conservation farming involves the use of those farming systems that sustain economic returns while preserving productive potential of the soil resource. Intensive cropping systems based on monoculture, intensive grazing with a high stocking rate and unscientific agronomic practices lead to severe erosion. Using diverse farming systems, based on a judicious mix of crops, trees, and livestock, can minimize risks of soil erosion. In Turkey, for example, Mete (1976) observed that natural grasslands, vetch, and oats–cotton rotation caused the minimum soil erosion. He recommended a liberal use of farmyard manure every year to maintain a good soil structure. In the Yemen Arab Republic, Alkamper *et al.* (1979) recommended that agricultural improvements be sought in areas of animal husbandry and afforestation. They also recommended that restrictions be placed on absentee graziers by providing them with suitable employment opportunities.

Proper selection of crops, rotation, and crop management systems are important agronomic considerations for erosion control. Grain crops should be grown in rotation with planted fallows. The fallow can be a leguminous cover or a lightly grazed pasture. Mixed cropping of cereals and legumes may be more conservation effective than monocropping. Mixed cropping may be organized into contour strip cropping with alternate strips planted to grain crops and pastures. Strip boundaries can be made permanent by establishing perennial hedges. These hedges can serve as wind breaks, barriers to reduce runoff velocity, and as source of fodder and fuelwood. The choice of species for contour hedgerows, however, is an important consideration. The management of hedgerows can be labour-intensive.

Tillage   Tillage, physical soil manipulation to create optimal environment for seed germination and seedling establishment, plays a significant role in soil and water conservation. There are a wide range of tillage tools, and different systems of tillage can be created by combining the type of tillage implements with frequency, intensity, and timing of their use (Table 6.11). Depending upon antecedent soil properties and the method/type of tillage used, tillage may increase or decrease the risk of soil erosion.

At the onset of rain, most soils of semi-arid and arid regions are naturally compacted and have low infiltrability. Runoff losses from a single storm can

Table 6.11   Tillage practices for erosion control

Objective	Tillage system
1. Breaking surface crust	Shallow ploughing
2. Improving infiltration	Chiselling, sub-soiling
3. Increasing surface detention	Mouldboard ploughing, pitting, plough up at the end of rains
4. Reducing runoff	Contour cultivation, ridge–furrow system, tied-ridges or basin tillage
5. Decreasing soil erosion while improving drainage	Graded beds and furrows, land smoothing, camberbed technique

exceed 50% of the rain received, especially when rainfall intensity exceeds infiltration capacity of the soil. Reduced infiltration capacity is often caused by the development of surface crust, and the presence of a natural or cultivation-induced sub-surface horizon of restricted permeability to water. The objectives of tillage should, therefore, be to break the crust, increase surface detention capacity, and improve water-transmission characteristics of the sub-soil horizon. On the basis of soil and climatic constraints to crop production, Lal (1985) developed a soil guide to tillage needs (Figures 6.5 and 6.6). Appropriate tillage methods to alleviate soil-related constraints and to reduce erosion hazards are described below:

(a) *Rough and cloddy seedbed:* A rough seedbed with a cloddy and uneven surface is more suited to decreasing the risk of crust formation and improve the surface detention capacity than a fine seedbed with a smooth surface. Intensive surface management studies have been conducted in semi-arid and arid regions of West Africa (Charreau, 1972, 1977; Charreau and Nicou, 1971; Chopart and Nicou, 1976; Chopart, 1981). Most studies indicate the benefits of a rough and cloddy seedbed in terms of decreasing losses due to water runoff and conserving more water for plant growth. Chopart *et al.* (1981), while summarizing the results from tillage studies conducted in Senegal, Togo, and Ivory Coast, concluded that structurally inert soils of the arid regions benefit markedly from mechanical tillage. Similar conclusions were arrived at for soils of the semi-arid and arid regions of Botswana by Willcocks (1981). These soils are mostly sandy, and are underlain by a thick ferruginous concretionary horizon. Periodic sub-soiling to improve the infiltration rate of the compacted sub-soil is, therefore, necessary to reduce runoff. An early tillage that creates an erosion-resistant rough soil surface is useful (Spath, 1975). Tillage performed at the end of the rainy season or that performed early enough to provide a rough and cloddy soil surface are

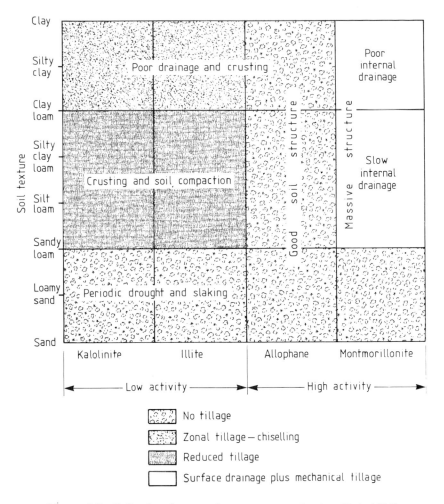

Figure 6.5   Soil-related constraints to crop production (Lal, 1985)

more conservation effective than those tillage systems that render the soil surface smooth and devoid of vegetation cover (Nicou, 1974a, 1974b).

Disc ploughing and harrowing to create a smooth seedbed must be avoided. These tillage implements may destroy aggregates and loosen the soil particles for an easy transport with overland flow. Disced soil is also prone to developing crust and surface seal. Experiments conducted in southern Tunisia have indicated the adverse effects of disc ploughing on soil erosion. Floret and Le Floch (1984) reported that disc ploughing

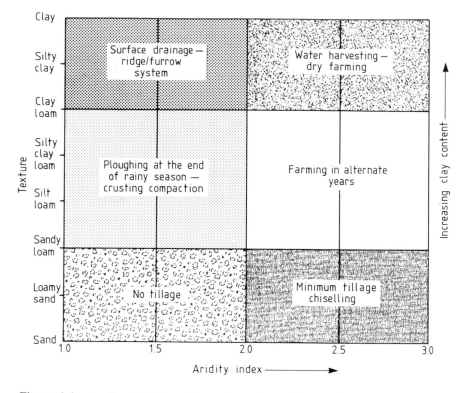

Figure 6.6   A soil suitability guide to appropriate tillage methods (from Lal, 1987).

caused erosion of 8 to 12 mm of soil per year. Excessive tillage performed by multiple tillage operations must also be avoided.

Research experiments conducted by ACSAD in Syria have provided specific tillage recommendations for erosion control in arid regions (Zagit-Az, 1978). These recommendations are as follows:

- contour ploughing with steep-walled furrows that are deep enough to retain and absorb heavy runoff,
- furrows to be sown to economically useful plants that also have an extensive root system to bind the soil, and
- efforts to be made to suppress rill formation.

Rill formation can be suppressed by contouring and a series of ridge–furrow systems. 'Contouring' involves ploughing, planting, cultivation, etc., across the slope of the land so that elevations along rows are as near to level as possible. The contour ridge system can be strengthened against breakage by tie-ridging.

(b) *Tied-ridge system or basin tillage:* Tied-ridging or basin tillage is contour ridges with the addition of cross ties in the furrows to hold surplus water and allow more time for infiltration into the soil. Tied ridges have proven effective in soil and water conservation in a wide range of soil and ecological environments (Lal, 1989; Hulugalle, 1987). The data by Rawitz *et al.* (1983) from Israel showed that erosion loss from disced and ploughed plots was approximately 10 times more than that from the ridged and 25 times more than that from the tied-ridge systems. Also in Israel, Agassi *et al.* (1986) reported very high efficiency of the basin-tillage method for soil conservation in cotton. Their data showed that basin tillage yielded only 10% of the runoff of the ridged plots (5.6% versus 60% of the rainfall) (Figure 6.7). Erosion rates and accumulated erosion were even less, 1.5 t/ha in comparison with 25 t/ha in the ridged-and-furrow plots, respectively (Figure 6.8). Agassi *et al.* (1986) made the following recommendations for improving the efficiency of the basin tillage system:

- The basin should not be isolated from the soil beds and the slightly convexed bed should always lead the runoff water to the basin.
- The dyke crust should always be higher than the basin's shoulder in order to prevent water overflow from one basin to another.

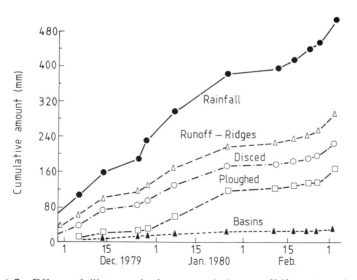

Figure 6.7   Effects of tillage methods on cumulative runoff (from Agassi *et al.*, 1986)

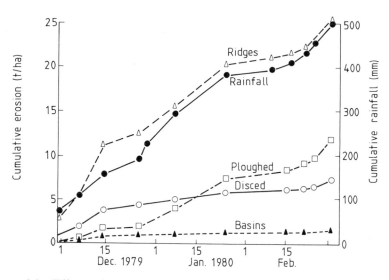

Figure 6.8   Effects of tillage methods on soil erosion (from Agassi *et al.*, 1986)

Another technique for increasing surface storage is to develop a large number of surface pits at frequent intervals. Similar to basin tillage, pitted plots also have a lower runoff and less soil erosion. The pitted seedbed is widely used in the Sahel and in the sub-Saharan region.

(c) *Mulching:* Using crop residue and other biomass at the soil/air interface is another useful practice to reduce raindrop impact, decrease surface crusting, improve infiltration, and reduce runoff and soil erosion. Mulching enhances activity of soil fauna and improves macroporosity. Channels and feeding galleries created in a termite-infested soil increase infiltration and water transmission (Lal, 1987). The infiltration rate is improved if termite activity increases macropores and encourages formation of large-size water stable aggregates. Otherwise, termite activity may also decrease infiltration. In Texas, Spears *et al.* (1975) reported lower infiltration on termite-infested than on termite-free soil because of a high soil organic matter content in the latter. Williams (1979) observed from Gambia that water infiltration into the soil of an old mound surface was very slow. Crop residues are scarce in arid climates. The residues are used as animal feeds, for fencing, and for fuel, and are not available for mulching. The practice known as 'stubble mulch farming' can be useful provided that an adequate amount of residue mulch is left on the soil surface. Because the soil surface is often crusted and compacted, surface tillage to break the crust without incorporating crop residue can be a beneficial practice. On an Austin clay in Texas, with a 4% slope and rainfall of 216 mm in one day, Adams (1966) observed significant reduc-

Table 6.12 Infiltration, runoff, and erosion under different mulches on an Austin clay on 4% slope in Texas, USA. Rainfall = 216 mm in one day

Mulch treatment	Infiltration (mm)	Runoff (mm)	Erosion (t/ha)
Bare control	35.8	179.8	38.0
5-cm straw mulch (20 t/ha)	53.3	162.3	Traces
2.5-cm gravel mulch	84.1	131.6	Traces
5-cm gravel mulch	123.4	92.2	Traces

Modified from Adams (1966).

tion in runoff and erosion by mulching with straw or gravel (Table 6.12). In southern Tunisia, Floret and Le Floch (1984) observed that mulching with barley straw and with the twigs of *Aristida pungens*, *Rhantherium suaveolans*, and *Artemisia campestris* drastically reduced water erosion. The use of alternating a vegetation strip with cultivation strips and incorporation of crop residue in the plough layer were also useful in reducing runoff and erosion. In Iran, Hakimi *et al.* (1976) also recommended vertical mulching to conserve water by reducing losses due to runoff. Vertical mulching is a technique of inserting crop residue into narrow trenches dug on the contour.

Because crop residue is not easily available, many supporting practices are recommended to procure crop residue mulch. Some cover crops or forages are specifically grown to procure mulch. Some important economically useful species include *Trifolium* spp. and *Medicago* spp. Checker board planting is another useful technique (FAO, 1984). Dead palm branches are used to demarcate small squares. Drought-resistant grass and perennial shrubs are planted within these squares to conserve soil and water resources.

(d) *Soil fertility management:* Poor vegetation cover in soils of the arid regions is partly due to inherently low soil fertility. Soils of these regions are often low in organic matter content and in essential plant nutrients. These soils may contain sufficient quantities or even excess of some salts of Na^+, K^+, Ca^{2+} and Mg^{2+} but are deficient in essential nutrients such as N, P, Zn, etc. Improving soil fertility, increasing soil organic matter content, and creating a balanced status of essential nutrients may be essential prerequisites for good farming. Low crop stand, poor and stunted plant growth, and low canopy cover enhance soil erosion. A good crop growth through balanced application of plant nutrients and improved soil fertility are important and effective soil-conserving measures. Intensive use of inorganic fertilizers, however, should be limited to those soils with an assured water supply, e.g. through supplementary irrigation for soils with a perpetual water deficit. For dry farming use of organic manures should be preferred over that of inorganic fertilizers.

High biomass production by improved soil fertility also leads to an increase in soil organic matter content. The latter can also be increased by application of regular and liberal doses of farmyard manure and other organic wastes. Increase in soil organic matter content improves soil structure and increases the soil's capacity to store plant-available water reserves. High soil organic matter content also increases biomass carbon and enhances microbiological activity (Toutain, 1974).

(e) *Soil conditioners and amendments:* Aggregation and water stability of aggregates can be improved by the application of some soil conditioners. Farmyard manure and other organic wastes also act as soil conditioners. These synthetic or natural polymers bind the soil particles into domains, microaggregates, and aggregates. Well-aggregated soils, with a predominance of micro- or macroaggregates, are less prone to erosion by water. In addition to synthetic conditioners, bitumen or asphalt has also been used in soils of arid regions to increase aggregation of coarse single-grained particles. In Tunisia, De Kesel and De Vleeschauwer (1981) advocated the use of polyurea polyalkylene oxide (Uresol) for improving aggregation of very loose soils. They observed that sand treated with Uresol become resistant to water erosion, and tree seedlings established faster on stabilized soil than on an untreated control.

Synthetic soil conditioners and other capital-intensive measures may be useful in special circumstances. Sub-surface barriers of asphalt, plastic film, or bentonite clays have been used in arid regions for conserving water in the root zone (Erickson *et al.*, 1968). Sub-surface moisture barriers are used to improve water storage in the root zone of sandy desert soils. These barriers can be combined with other techniques of trenching or microcatchments to concentrate runoff. The moisture-barrier-runoff concentration technique was found to be successful in

Table 6.13 Soil moisture storage (mm) in the top 60 cm by the use of moisture barrier-runoff concentration techniques, Jodhpur, India

Treatment	1 July	27 July	9 August	19 August
		Date of sampling in 1977		
Barrier-runoff concentration	107	100	117	61
Control	80	72	91	54

Each value is an average of two separate measurements for the trench and the pit. Modified from Singh (1976).

reducing runoff and increasing soil-water storage in sandy soil of an arid region near Jodhpur, India. The data in Table 6.13 show a significant increase in water storage in the root zone by this method (Singh, 1976).

However, soil conditioners and these special techniques may be uneconomical for general use on arable lands or in pastures.

Erosion Control Measures

Runoff management and safe disposal of surplus water are the basic principles of erosion control measures. Runoff management involves (a) prevention of runon from outside the land, (b) decreasing velocity of surplus runoff from the land, and (c) storage of runoff in appropriately sited tanks for supplementary irrigation. These three objectives are achievable by installing engineering structures. Engineering structures, however, have some severe limitations. They are expensive to install and maintain and require special skill. These devices cannot be used for soil and water conservation on small farms in many countries. Some of the commonly used techniques on large commericial farms are listed in Table 6.14.

Water harvesting The objective of water harvesting is to store surplus runoff for productive use and to trap the silt carried. Water harvesting involves treating unproductive/marginal uplands to encourage runoff and its collection in specially designed tanks for eventual use as supplementary irrigation. This is an ancient practice and has been used in Negev Highland Desert since about 1000 BC (Unger, 1984).

Some commonly used water harvesting techniques, especially those used in arid regions of Australia, are described by Hollick (1976). Runoff from the surrounding hills should be collected with a great caution. Watersheds should

Table 6.14 Engineering techniques of runoff management for erosion control

Objectives	Engineering techniques
1. Preventing runon	Diversion ditches, storm water drains
2. Reducing runoff velocity	Land forming/levelling, graded channel terraces, drop structures, chutes, grass waterways
3. Gully control	Gabions, drop structures, check dams
4. Supplementary irrigation	Water harvesting through graded furrows/beds, grass waterways, storage dams

be so treated as to prevent/minimize erosion. This is one of those special situations where steep hill slope can be treated with asphalt/bitumen emulsions, plastic films, or even butyl rubber to encourage runoff without causing soil erosion. Soil can also be sprayed with water repellents. Rock rip-rap and stones are also recommended for controlling erosion on steep ridges.

Preventing runon   Water from upslope areas should be diverted away from the farmland by installing diversion ditches or storm-water drains. Diversion ditches are constructed across the slope to intercept the runon and divert it to a point where it can be stored without causing erosion en route. Vegetated waterways, wide and shallow and protected by good grass and other vegetation cover, can be used to divert the water from hillslopes of gentle gradient (<15%). Grass cannot be successfully grown on very steep slopes. For slopes exceeding 15%, waterways are often paved with stones. Stoned waterways can also serve as footpaths.

Reducing velocity of runoff from cropland   Different types of terraces are recommended for reducing velocity of overland flow and for safe disposal of surplus runoff. Commonly used terraces are graded channel terraces, level terraces with open or closed ends, bench terraces, broad terraces, etc. Bench terraces can have a front wall protected by stones or grass. These terraces are mostly used for growing fruit trees.

Terrace agriculture is suitable for those steeplands where frequent maintenance is possible. However, soil erosion can be more severe from improperly constructed and poorly maintained terraced than unterraced lands. Fruit terraces are widely used on steeplands in Jordan in regions where 70% of the land receives about 250 mm of rain a year (FAO, 1984). With well-constructed and properly maintained terraces, those slopes are being used for production of olives, peaches, almonds, pomegranates, and grapes. Terrace faces are protected by stone walls. These slopes are not suitable for cultivation of grain crops because of excessive erosion. According to some estimates, unterraced slopes used for grain crop production have lost most of the topsoil in the Mediterranean Basin.

Installing a contour bank is another technique used to reduce runoff velocity on gentle slopes. Earthen banks are widely used in arid and semi-arid regions of India. Contour banks can be useful, especially if designed with a graded channel to dispose of excess water at a gentle velocity. Contour banks can be stabilized by growing suitable trees or shrubs.

Another variant of an earthen dam is the rock bund. Rock bunds are recommended for reducing runoff velocity, increasing infiltration, and encouraging sedimentation on gently sloping lands in arid regions of Africa.

Table 6.15 Controlling runoff and soil erosion by rock bunds in arid regions of Niger

Management system	Runoff (% of rainfall)	Soil erosion (t/ha yr)
1. Traditional farming by hoe cultivation	17.6	10.1
2. Buffer strip cropping—ploughing, ridging, weeding	5.2	1.2
3. Stone lines 80 cm high	3.8	0.5
4. Earth dykes with stones	0.9	0.3

Modified from Roose (1977).

Porous rock bunds improve water infiltration and increase the probability of establishment and survival of young seedlings of weeds and shrubs. The data in Table 6.15 show significant reduction in runoff and erosion in the Allokoto region of Niger by the installation of rock bunds.

Controlling gully erosion Gully control can be achieved by a range of techniques such as check dams, gabions, and drop structures. Installation of a diversion channel is an important prerequisite for restoration of a large gully. Once the excess water is diverted away from the gully, check dams and drop structures can be established to restore the gully. Check dams involve construction of a succession of small dams along the hillslope to retard the runoff velocity and to encourage sedimentation. These dams slow down the velocity of water runoff and allow more time for water to infiltrate into the soil. The dam size successively increases downstream as the valley widens. Large dams are often provided with a stone rip-rap and an overflow system to safeguard against floods. Fruit trees and other useful perennials can be grown around the dam site to provide protective vegetation cover and to stabilize the dam. In southern Tunisia, a commonly recommended plant to stabilize these dams is *Stipa tenacissima*. Gabions are check dams of loose stone held together by a wire mesh.

Drop structures are specially designed engineering structures to reduce the runoff velocity in a rapidly developing gully. There are two types of drop structures: low drop structures and high drop structures (Blaisdell, 1981). A low drop structure is defined as a drop in which the bed drop height is equal to or less than the upstream specific head, that is less than the upstream flow depth plus the velocity head. High drop structures are those in which the upstream water levels are normally unaffected by downstream conditions, such as high tailwater levels that submerge the spillway crest.

Chutes are devices designed to safely discharge runoff on steep gradients. There are two types of chutes for erosion control (Blaisdell, 1981). A plain chute is used to transport the flow at high velocity with minimum energy dissipation. A baffled chute is used to dissipate the energy as fast as it is generated by the chute slope.

RESTORATION OF ERODED LANDS

Soil resources of arid regions are not only finite but also non-renewable at the human timescale. Restoration of eroded and degraded land is, therefore, a high priority. An important step towards restoring a degraded ecosystem is afforestation. Afforestation programmes should be implemented for all high-lands and steeplands that have been denuded of their vegetation cover by overgrazing and expansion of agriculture to marginal lands. Establishing tree seedlings can be difficult, however, in these harsh environments. Arido-active species are more suitable/adaptable to xerophytic conditions than arido-passive species (Floret and Le Floch, 1984). Arido-active species have better water-use efficiency than arido-passive species. Some useful plants adaptable for different ecological regions of Africa are listed in Table 6.16. Useful arid zone plants recommended for southern Tunisia are *Artemisa herba-alba*, *Helianthemum lippii*, *Stipa lagascae*, *Rhantherium suaveolens*, *Argyrolobium uniflorum*, and *Echiochilon* spp. (Floret, 1981). These are perennial pastoral plants and can also withstand a normal grazing pressure. In Egypt and other arid regions *Casurina*, *Eucalyptus*, and *Cypress* trees are suitable for soil conservation and as windbreaks. Seedling establishment on eroded and degraded lands can be improved by a combination of rock bunds and microcatchments. This practice is being used to establish trees on degraded lands in Niger.

Table 6.16   Useful tree species adapted for arid regions of Africa

Region	Tree species
1. Sahelian	*Acacia albida*, *A. seyal*, *A. raddiana*, *A. senegal*, *A. laeta*, *A. ehrenbergiana*, *A. nilotica*, *A. Sibriana*, *Commiphora africana*, *Balanites aegyptica*, *Salvadoria persica*, *Maerue crassifolia*, *Calotropis procera*, *Euphorbia balsamifera*
2. Sudanian semi-arid	*Azadirachta indica*, *Hyphaena thebaica*, *Sclerocarya birrea*, *Adansonia digitata*, *Parkia* sp., *Acacia albida*, *Tamarandus indica*, *Bombax costatum*, *Anogeissus* spp., *Balanites* spp., *Prosopis* spp., *Butyrospermum* spp.

SUMMARY AND CONCLUSIONS

Water erosion is a problem in most arid and semi-arid regions. Susceptibility of soils to erosion is attributed to low ground cover and denuded landscape, high-intensity and torrential rains, surface seal and crust leading to low infiltration, and high erodibility due to low levels of soil organic matter content. Soil erosion assessment can be made by visual observations establishing field runoff plots and through modelling. Field measurements must be made by standardized technologies and for a long time period of at least 3 years to establish long-term trends.

Erosion control on agricultural lands can be achieved by soil and crop management systems that ensure a good cover close to the ground surface. Crop residue mulch, sod crops, and multiple cropping systems are effective in reducing inter-rill/rill erosion and soil splash. Slope length plays an important role in concentrating runoff. Slope gradient and length can be reduced by installing terraces or dams at intervals across the slope. There are many kinds of terraces to meet the specific needs. Establishment and maintenance of good vegetative cover, however, is the best way to control erosion.

The following conclusions can be drawn from the literature surveyed in this report:

(a) Change in land use, denudation of vegetation cover, torrential rains and flash floods are the causative factors responsible for severe water erosion hazard in arid lands.

(b) There is a scarcity of experimental data on the effectiveness of different measures in controlling runoff and erosion in soils of arid regions.

(c) Erosion preventive measures are to be preferred over the control measures. The latter are expensive, usually less effective, and often too late to be useful.

(d) Appropriate land use, suitable crop rotations, judicious tillage methods, fertility maintenance, and other agronomic practices of 'good farming' are useful and cost-effective erosion preventive measures.

(e) Terraces and other engineering techniques can be useful, especially if properly constructed and adequately maintained.

(f) Restoring eroded and degraded ecosystems is a high priority, and afforestation can be successful if tree species are chosen for special ecosystems.

REFERENCES

Adams, J. E. (1966). Influence of mulches on runoff, erosion and soil moisture depletion. *Proc. Soil Sci. Soc. Am.*, **30**, 110–16.

Agassi, M., Benyamini, Y.. Morin, J., Marish, S. and Henkin, A. (1986). *The Israeli concept for runoff and erosion control in semi-arid and arid zones in the Mediterranean basin*, Soil Erosion Research Station, Rupin Institute Post, Emer-Hefer, Israel, 76 pp.

Alkamper, J., Hauffuer, W., Matter, H. E., Weise, O. R. and Weiter, M. (1979). Erosion control and afforestation in Haraz, Yemen Arab Republic, *Gussener Beitrage Zur Enwicklungsforschung, Reihe*, **II**(2), 106 pp.

Balci, A. N. (1973). Influence of parent material and slope exposure on properties of soils related to erodibility in north-central Anatolia, Fakultesi Yayinlari, Istanbul Universitesi, Turkey, No. 195, 60 pp.

Balci, A. N. and Ozyuvaci, N. (1974). Variation in erodibility of soils as related to parent materials, slope exposure, land use and sampling depth in two different regions of Turkey, *Faculty of Forestry, Istanbul University, Buyukdere, Istanbul, Turkey*, **24**(2), 79–107.

Blaisdell, F.W. (1981). Engineering structures for erosion control, in *Tropical Agricultural Hydrology* (Eds. R. Lal and Z. W. Russell), pp. 325–55, J. Wiley & Sons, Chichester.

Bonvallot, J. and Hamza, A. (1977). Causes and modelities of erosion in the lower basin of the Wad El-Hadjel (Central Tunisia), IAHS Publication 122, ORSTOM, 18 Avenue Charles, Nicdle, Tunis, Tunisia, pp. 260–8.

Boukhobza, M. (1982). Pastoral farming and erosion in the high Plateaux steppe lands of Algeria, Production Pastorale et Société, Ministère Algérien de la Planification et de l'Aménagement du Territoire, Algiers, Algeria, No. 10L64–67.

Budyko, M. I. (1974). *Climate and Life*, Academic Press, New York, 508 pp.

Chaabouni, A. (1977). A study of solid transport under a traditional management program using surface runoff to nourish olive plantations in the Sahel region of Tunisia, IAHS Publication 122, pp. 284–91.

Charreau, C. (1972). Problèmes posés par l'utilisation agricole des sols tropicaux par des cultures annuelles, *L'Agr. Trop.*, **27**, 905–29.

Charreau, C. (1977). Some controversial technical aspects of farming sytems in semi-arid West Africa, in *Proceedings of International Symposium on Rainfed Agriculture in Semi-arid Regions* (Ed. G. H. Cannell), pp. 313–60, University of California, Riverside, Calif.

Charreau, C. and Nicou, R. (1971). L'amelioration du profil cultural dans les sols sableux et sables argileux de la zone tropicale sèche Ouest Africaine et ses incidences agronomiques, *L'Agro. Trop.*, **26**, 209–55, 565–631, 903–78, 1184–247.

Chopart, J. L. (1987). *Le Travail du Sol au Sénégal*, ISRA, Bambey, Senegal.

Chopart, J. L. and Nicou, R. (1976). Influence du labour sur le developpement radiculaire de differentes plantes cultivées au Senegal—Consequences sur leur alimentation hydrique, *L'Agro. Trop.*, **31**, 7–28.

Chopart, J. L., Kalms, J. M., Marquette, J. and Nicou, R. (1981). *Comparison de Differentes Techniques de Travail du Sol en Trois Écologies de l'Agrique de l'Ouest*, IRAT, Montpellier, France.

De Kesel, M. and De Vleeschauwer, D. (1981). Sand dune fixation in Tunisia by means of polyurea polyalkylene oxide (Uresol), in *Tropical Agricultural Hydrology* (Eds. R. Lal and E. W. Russell), pp. 273–81, J. Wiley & Sons, Chichester.

Dregne, H. (1976). *Soils of Arid Regions*, Elsevier Scientific Publishing Co., Amsterdam, 237 pp.

Elwell, H. A. and Stocking, M. A. (1975). Parameters for estimating annual runoff and soil loss from agricultural lands in Rhodesia, *Water Resources Res.*, **11**(4), 601–5.

Erickson, A. E., Hanson, C. M. and Smucker, A. J. M. (1968). The influence of sub-surface asphalt moisture barriers on the water properties and productivity of sand soils, *Trans. 9th Int. Congr. Soil Sci. Adelaide*, **I**, 331–7.

FAO (1979). A provisional methodology for soil degradation assessment, FAO/

UNEP/UNESCO Report Food and Agricultural Organization of the United Nations, Rome, Italy, 84 pp.

FAO (1984) *Protect and Produce*, FAO, Rome, Italy, 40 pp.

Finkel, H. J. and Finkel, M. (1986). Hydrology, in *Semi-arid Soil and Water Conservation* (Eds. H. J. Finkel, M. Finkel and Z. Naveh), pp. 5–26, CRC Press, Inc., Boca Roton, Fla.

Floret, C. (1981). The effects of protection on steppic vegetation of the mediterranean arid zone of Southern Tunisia, *Symposium on Dynamique de la végétation dans les formations mediterranéennes, lingneeses*, Montpellier, 15–20 Sept. 1980; *Vegetation*, **46**, 117–29.

Floret, C. and Le Floch, E. (1984). Agriculture and desertification in arid zones of northern Africa, UNEP/USSR Workshop on Impact of Agricultural Practices, Batumi, USSR, October 1984, Mimeo, 18 pp.

Foster, G. R. (1982). Modelling the soil erosion process, in *Hydrologic Modelling of Small Watersheds*, ASCE, New York, Preprint 82-007.

Foster, G. R. (1988). Modeling soil erosion and sediment yield, in *Soil Erosion Research Methods* (Ed. R. Lal), pp. 97–118, Soil and Water Conservation Society, Ankeny, Iowa.

Foster, G. R. and Meyer, L. D. (1975). Mathematical simulation of upland erosion by fundamental erosion mechanics, in *Present and Perspective Technology for Predicting Sediment Yields and Sources*, ARS-S40, pp. 190–207, USDA, Washington.

Foster, G. R., Meyer, L. D. and Onstad, C. A. (1977). A runoff erosivity factor and variable slope length exponents for soil loss estimates, *Trans. ASAE*, **20**, 683–7.

Gerrard, A. J. (1981). *Soils and Land Forms*, George Allen and Unwin, London, 219 pp.

Ghazi, A. E. M. (1982). The influence of soil structure and fineness of $CaCO_3$ on water movement in highly calcareous soils, *Egyptian Journal of Soil Science (Egypt)*, **22**(2), 143–54.

Goudie, A. and Wilkinson, J. (1977). *The Warm Desert Environment*, Cambridge University Press, Cambridge, UK, 88 pp.

Haan, C. T., Johnson, H. P. and Brakensiek, D. L. (Eds.) (1982). *Hydrologic Modeling of Small Watersheds*, ASAE, St Joseph, Mich.

Hakimi, A. H., Kachru, R. P. and Chakrabarti, S. M. (1976). *Tillage in Difficult Soils: Dryland Farming in Iran*, Vol. 28(1), pp. 8–10, College of Agriculture, Pahlavi University, Shiraz, Iran.

Hamza, A. and El-Amami, M. (1977). The use of satellite imagery in morphopedological mapping, as exemplified by the map of Sbeitla 1:200000. Organized by International Society of Soil Science, Commission V. (Ed. M. C. Girard), Tunis, Tunisia, pp. 125–37.

Hollick, M. (1976). Water harvesting in Australia, in *Arid Zone Research and Development* (Ed. H. S. Mann), pp. 269–80, Scientific Publishers, Jodhpur.

Hudson, N. W. (1965). The influence of rainfall on the mechanics of soil erosion with particular reference to Southern Rhodesia, M.Sc. Thesis, University of Cape Town.

Hulugalle, N. R. (1987). Effects of tied ridges on soil-water content, evapotranspiration, root growth and yield of cowpeas in the Sudan Savanna of Burkina Faso, *Field Crops Res.*, **17**, 219–28.

Kinnell, P. I. (1981). Rainfall intensity-kinetic energy relationships for soil loss prediction, *Soil Sci. Soc. Am. Proc.*, **45**, 153–5.

Kirkby, J. J. (1971). Hillslope-process-response models based on the continuity equation, in *Slopes Forms and Processes* (Ed. D. Brunsden), pp. 15–30, Institute of British Geographers Special Publication 3.

Knisell, W. G. (Ed.) (1980). CREAMS: a field scale model for chemicals, runoff and erosion from agricultural management sytems, USDA Conservation Research Report 26, 640 pp.

Kramer, L. A. and Meyer, L. D. (1969). Small amount of surface mulch reduce soil erosion and runoff velocity, *Trans. ASAE*, **12**, 638–41, 645.

Labib, T. M. (1981). Soil erosion and total denudation due to flash floods in the Egyptian eastern desert, *Journal of Arid Environments*, **4**(3), 191–202.

Laflen, J. M., Baker, J. L., Hartwig, R. O., Buchelle, W. F. and Johnson, H. P. (1978). Soil and water loss from conservation tillage system, *Trans. ASAE*,**21**, 881–6.

Lal, R. (1976). Soil erosion problems on a tropical Alfisol and their control, IITA Monograph 1, Ibadan, Nigeria, 208 pp.

Lal, R. (1985). A soil suitability guide for different tillage methods in the tropics, *Soil and Tillage Res.*, **5**, 179–98.

Lal. R. (1987). *Tropical Ecology and Physical Edaphology*, J. Wiley & Sons, Chichester, UK.

Lal, R. (1988a). Effects of slope length, slope gradient, tillage methods and cropping systems on runoff and soil erosion on a tropical Alfisol: preliminary results, *IAHS Publ.*, **174**, 79–88.

Lal, R. (1988b). *Soil Erosion Research Methods*, Soil and Water Conservation Society, Ankeny, Iowa, 244 pp.

Lal, R. (1989). Conservation tillage for sustainable agriculture: tropics vs. temperate environments, *Adv. Agron.*, **43** (in press).

Le Houérou, H. N. (1970). North Africa: past, present, and future, in *Arid Lands in Transition* (Ed. H. E. Dregne), pp. 227–78, AAAS, Washington, DC.

Matlock, W. G. (1981). *Realistic Planning for Arid Lands: Advances in Desert and Arid Land Technology and Development*, Harwood Academic Publishers, New York, 261 pp.

Meyer, L. D. (1988). Rainfall simulators for soil conservation research, in *Soil Erosion Methods* (Ed. R. Lal), pp. 75–96, Soil and Water Conservation Society, Ankeny, Iowa.

Meyer, L. D. and Monke, E. J. (1965). Mechanics of soil erosion by rainfall and overland flow, *Trans. ASAE*, **8**, 572–7, 580.

Mete, C. (1976). Soil and water losses in Cukurora transitional lands as influenced by organic matter, top- and sub-soil conditions and cotton farming systems applied on bare and covered soils, Ziraat Yuksek Muhendisi, Tarsus Boloe Topsaksu Arastiruna Enstitusu, Tarsus, Turkey, Genel Yayin, No. 73, Rap. Ser. No. 28, 95 pp.

Monjauze, A. (1960). Solutions doctrinales du problème pastoral dans les régions a climats xèrothermiques, in *Compete Rendu et la Restauration des sols*, pp. 307–19, Institut Français de Cooperation Technique et Faculté d'Agronomie de l'Université de Teheran.

Morgan, R. P. C. (1978). Field studies of rain splash erosion, *Earth Surface Processes*, **3**, 295–9.

Mutchler, C. K. and Greer, J. D. (1980). Effect of slope length on erosion from low slopes, *Trans. ASAE*, **23**, 866–72.

Mutchler, C. K. and Larson, C. L. (1971). Splash amounts from water drop impact on a smooth surface, *Water Resources Res.*, **7**, 195–200.

Mutchler, C. K., Murphree, C. E. and McGregor, C. K. (1988). Laboratory and field plots for soil erosion studies, in *Soil Erosion Research Methods* (Ed. R. Lal), pp. 9–38, Soil and Water Conservation Society, Ankeny, Iowa.

Nicou, R. (1974a). Contribution to the study and improvement of the porosity of sandy-clay soils in the dry tropical zone: agricultural consequences, *Agro. Trop.*, **29**, 1110–27.

Nicou, R. (1974b). The problem of caking with the drying out of sandy and sandy clay soils in the arid tropical zone, *Agr. Trop.*, **30**, 325–43.

Nir, D. (1974). *The Semi-arid World: Man on the Fringe of the Desert*, Longman, 187 pp.

Palmer, R. S. (1963). The influence of a thin water layer on water drop impact forces, *Int. Assoc. Sci. Hydrology Bull.*, **65**, 141–8.

Rawitz, E., Morin, J., Hoogmoed, W. B., Margolin, M. and Etkin, H. (1983). Tillage practices for soil and water conservation in the semi-arid zone. I. Management of fallow during the rainy season preceeding cotton, *Soil and Tillage Res.*, **3**, 211–31.

Renard, K. G. (1982). Comments on 'Soil erosion and total denudation due to flash floods in the Egyptian desert', *Journal of Arid Environments*, **5**(4), 347–51.

Ritchie, J. C., Hawks, P. H. and McHenry, J. R. (1974). Estimating soil erosion from the redistribution of fallout ^{137}Cs, *Soil Sci. Soc. Am. Proc.*, **38**, 137–9.

Ritchie, J. C., Hawkes, P. H. and McHenry, J. R. (1975). Deposition rates in valley determined using fallout caesium-137, *Geol. Soc. Am. Bull.*, **86**, 1128–30.

Roose, E. J. (1977). Adaption of soil conservation techniques to the ecological and socio-economic conditions of West Africa, *L'Agron. Trop.*, **32**, 132–8.

Rose, C. W. (1988). Research progress on soil erosion processes and a basis for soil conservation practices, in *Soil Erosion Research Methods* (Ed. R. Lal), pp. 119–41, Soil and Water Conservation Society, Ankeny, Iowa.

Singh, H. P. (1976). Improving the moisture storage in sandy desert soils by sub-surface moisture barrier, in *Arid Zone Research and Development* (Ed. H. S. Mann), pp. 245–52, Scientific Publishers, Jodhpur.

Späth, H. J. (1975). Soil erosion and soil moisture balance in dry steppe climates with cold winters, as exemplified by central Anatolia, *Erdkunde*, **29**(2), 81–91.

Spears, B. M., Ueckert, D. N. and Whigham, T. L. (1975). Desert termite control in a short grass prairie: effect on soil physical properties, *Environ. Eng.*, **4**, 899–904.

Toutain, G. (1974). *Conservation of Soils in Saharian and Sahelian Date Palm Plantation*, Station Centrale d'Agronomie Sahariene (ORAD) Marrakech, Morocco.

Unger, P. W. (1984). Tillage systems for soil and water conservation, FAO Soils Bulletin 54, Rome, Italy, 278 pp.

USDA (1951). Soil Survey Manual, Agricultural Handbook 18, Government Printer, Washington.

Vanelslande, A., Lal, R. and Gabriels, D. (1987). The erodibility of some Nigerian soils: a comparison of rainfall simulator results with estimates obtained from the Wischmeier Nomogram, *Hydrological Processes*, **1**, 255–65.

Walker, P. H., Hutka, J., Moss, A. J. and Kinnell, P. I. A. (1977). Use of versatile experimental system for soil erosion studies, *Soil Sci. Soc. Am. J.*, **41**, 610–12.

Walling, D. E. (1988). Measuring sediment yield from river basins, in *Soil Erosion Research Methods* (Ed. R. Lal), Soil and Water Conservation Society, Ankeny, Iowa.

Willcocks, T. J. (1981). The tillage of clod farming sandy loam soils in the semi-arid climate of Botswana, *Soil and Tillage Res.*, **1**, 323–50.

Williams, J. B. (1979). Soil water investigation in the Gambia, Land Resources Development Center, ODA, UL Technical Bulletin 3, 183 pp.

Williams, J. R. (1975). Sediment yield prediction with universal equation using runoff energy factor, USDA-ARS-S 40, 244–52, Washington, DC.

Wischmeier, W. H. and Smith, D. D. (1958). Rainfall energy and its relationship to soil loss, *Trans. Am. Geophys. Un.,* **39**, 285–91.

Wischmeier, W. H., Johnson, W. B. and Cross, B. C. (1969). A soil erodibility nomograph for farmland and construction sites, *J. Soil Water Cons.*, **26**, 189–93.

Yair, A. and Lavee, H. (1985). Runoff generation in arid and semi-arid zones, in *Hydrological Forecasting* (Eds. M. G. Anderson and T. P. Burt), pp. 183–219, J. Wiley & Sons Ltd, UK.

Yair, A. and Rutin, J. (1981). Some aspects of the regional variation in the amount of available sediment produced by isopods and porcupines, northern Negev, Israel, *Earth Surface Processes and Landforms*, **6**(3–4), 221–34.

Yariv, S. (1976). Comments on the mechanism of soil detachment by rainfall, *Geoderma*, **15**, 393–9.

Young, R. A. and Wiersma, J. L. (1973). The role of raindrop impact in soil detachment and transport, *Water Resources Res.*, **9**, 1629–39.

Zagit-Az, M. (1978). Soil erosion in the coastal plains in the Arab Republic of Syria, Arab Center for the Studies of Arid Zones and Dry Lands, Report 17/78, Expert of Reforestation and Watersheds, Arabic Center, Damascus, Syria, 64 pp.

Zanchi, C. and Torri, D. (1980). Evaluation of rainfall energy in central Italy, in *Assessment of Erosion* (Eds. M. DeBoodt and D. Gabriels), pp. 133–42, Wiley.

CHAPTER 7

The Conservation and Use of Plant Resources in Dry Lands

J. Burley

Summary

Within the arid and semi-arid lands as defined throughout this volume natural vegetation types vary from xerophytic, often ephemeral, desert scrub and herbs through open savannah and savannah woodland towards deciduous forest. Land-use systems vary from nomadic and seasonal pastoralism to semi-permanent and permanent sedentary agriculture. Throughout these lands the pressures of human and domestic animal populations are increasing, land degradation is escalating and there is an urgent need to develop species and land-use systems that can meet human needs sustainably. Frequently, however, institutional factors are limiting; these include weakness of market, transport and communication systems, lack of trained and motivated staff, overall lack of political will, and the common perception of trees and shrubs as free goods not requiring deliberate interventions.

Trees and shrubs, whether natural or planted, offer a range of productive, environmental and social benefits. The quantitative effects are often difficult to assess (e.g. the role of tree roots in soil holding, water flow moderation and soil improvement, or the social value of diversity of products and risk reduction). The genetic resources of indigenous vegetation require exploration, evaluation, conservation and, for some species, genetic improvement; they can be used in a range of land-use systems including natural vegetation management and enrichment, industrial plantations (in wetter lands) and community or farm woodlots (in drier lands), and agroforestry (intimate mixtures of trees, shrubs, agricultural crops and possibly domestic animals).

Research is needed on choice of species and population, propagation and planting techniques, manipulation of soil microbial associates and recognition of human perceptions of plants and their values. A case study is described

Techniques for Desert Reclamation
Edited by A. S. Goudie
© 1990 John Wiley & Sons Ltd.

that covers the international testing of a number of species in different land-use systems that will be appropriate for drylands.

CHARACTERISTICS OF ARID AND SEMI-ARID LANDS

The climatic characteristics of dry lands were described earlier in this volume. Commonly used indices (such as that of UNESCO, 1979) take account of temperature, rainfall, radiation and wind effects, and this chapter is concerned with drylands in which plants can be economically and environmentally beneficial. In tropical Africa drylands account for more than 50% of the land surface (while supporting 35% of the human population) and in India some 127 million hectares of wasteland exist (degraded by deforestation and overcropping), most of which is in the dry classification.

Vegetation and Land-use Patterns

As one passes from the true desert towards more humid lands in Africa and India mean annual rainfall increases and the variation of mean annual rainfall decreases; the vegetation changes from xerophytic, often ephemeral, scrub and herbs, through open savannah and savannah woodland, towards deciduous and evergreen tropical forest.

Land use changes from nomadic pastoralism through seasonal transhumance to semi-permanent pastoralism and agriculture. In many countries government policies seek to encourage permanent agriculture even where shifting cultivation is the traditional land-use system, where human pressure on land is intensifying and where sustainability of crop production is not assured. The excessive use of land and the subsequent loss of productivity cause further desertification (see Chapter 1), but where shifting cultivation is the rule it is difficult to develop sufficient extension systems to teach and apply sedentary agricultural techniques.

In natural conditions and without excessive human and animal pressure, dryland ecosytems are able to maintain balanced exchanges of water and energy; however, removal of the vegetation exposes the soil surface to high temperatures and erosive forces so that organic material is mineralized, minerals are leached, soil structure is lost and entire soil surfaces are removed through erosion (see Chapters 3 and 6). Water percolation through the soil decreases with consequent effects on groundwater levels. In extreme cases soil erosion at one location causes the silting of streams, dams and irrigation systems downstream. Consequently there is an urgent need to develop land-use systems that reduce devegetation, species and land-use systems that provide all man's requirements and systems that rehabilitate degraded lands, whatever the cause of the original degradation.

Institutional Constraints

Throughout the drylands several institutional factors are common. Human populations are increasing with associated increases in the numbers of domestic animals, causing rapidly escalating rates of site degradation. Land tenure is predominantly communal with little incentive for the individual to make interventions that could enhance productivity and sustainability. Market infrastructures, communications and transport systems are weak, offering relatively little reason for the individual to produce more than subsistence quantities for himself and his family. Trained staff are not available for research, line management of forest or pasture, or for extension services that demonstrate appropriate systems to rural people, and where graduate staff are trained the drylands are commonly unpopular as a location for professional careers.

Overriding all these constraints is the traditional belief that trees and shrubs are free goods that do not require human intervention; only as shortages of common property resources become severe is the need for action recognized. In the case of drylands the basic common resource is the soil and when degradation is recognized it is often too late to take preventative action. There is thus an urgent need to initiate research and development programmes to determine optimum land-use systems that are sustainable and acceptable to local populations; this in turn requires the building or strengthening of institutions within the dryland countries.

BENEFITS OF TREES IN DRYLANDS

Productive Benefits

Trees and woody shrubs yield a wide range of products that are used for subsistence or sale in arid and semi-arid regions. The main groups include:

(a) building materials (poles for housing and posts for fencing and hedging),
(b) fuel (firewood and charcoal for domestic heating and cooking),
(c) fodder (leaves and fruits or pods for cattle),
(d) food (fruit and extractives for human food),
(e) fibre (clothing, ropes, thatch),
(f) chemical extractives (from leaves, bark and fruits or seeds for pharmaceuticals, adhesives, food additives, insecticides).

Environmental Benefits

Some of the environmental disbenefits of devegetation were indicated above. The positive benefits of trees and shrubs include the following main groups:

(a) reduction of soil erosion, loss of nutrients and structure, and damage to irrigation systems downstream (through root binding of soil and reduction of impact of raindrops on soil surfaces),
(b) reduction of water loss and frequency of flash flooding (by deep penetration of water into soil through root channels),
(c) improvement of soil structure and nutrient level (through nutrient pumping effect, mulching and nitrogen fixation in legumes with rhizobial associations and some other species having actinomycete associations, e.g. *Casuarina* species),
(d) improvement of microclimate, reduction of wind erosion of soil and provision of shelter for domestic animals and humans (by shelterbelts and trees around homesteads).

Many of these benefits are difficult to quantify, especially in financial terms, and many of them are seen as the perceived wisdom that is not necessarily proven. However, there is common consensus that, if the required minimum yield of productive benefits is maintained, the addition of trees and shrubs to land-use systems increases the diversity and sustainability of a given site.

Social Benefits

All the productive and environmental factors listed above clearly have social benefits to the individual, the family, the rural community and the nation. However, there are a number of other benefits commonly considered social that are harder to quantify yet are important elements in causing rural populations to plant or manage trees carefully. These include:

(a) diversity of subsistence diet or income from marketable products,
(b) reduction of risk through diversity of products and minimization of disease or inspect spread,
(c) maintenance of standing capital,
(d) counter-seasonality (the ability to use family or hired labour at times of low labour demand for agriculture or pasture management, and the ability to keep products standing until cash income is urgently needed or until market prices are favourable),
(e) reduction of artificial inputs such as fertilizer into sedentary agricultural land.

STAGES OF DEVELOPMENT OF PLANT GENETIC RESOURCES

There are two major groups of ways for sustaining and improving the yield of the benefit derived from plants. First, indigenous species may be improved genetically and traditional land-use systems improved managerially, possibly by small incremental changes. Second, exotic species may be identified, introduced and genetically improved while new land-use systems may be initiated. Underlying all systems that involve the planting of trees and shrubs are the accepted stages of the development of plant genetic resources that are adapted to the environmental, managerial and socal conditions of the locality.

Exploration

This involves the determination of the natural and introduced ranges of currently used or potentially useful species. Species that have wide natural ranges are subject to selection pressures exerted by different factors at different parts of the range so that population genetic structures vary. The major selection factors include temperature and rainfall regimes, particularly extremes, soil type and photoperiod; the exploration phase therefore seeks to map the total range of a species, relate it to the distribution of environmental factors and collect seed for comparative testing of the different populations at the evaluation stage.

Tree planting is not widely practised in dry regions and little experience of different species or populations within species exists. There is thus considerable opportunity for new exploration both of locally indigenous species and of exotics originating in other locations or countries. The cost and complexity of arranging explorations are great, particularly for species that occur naturally in several countries; because a given species may have potential value in many other countries, it is preferable to arrange a single, coordinated programme of exploration and evaluation.

Programmes for species suitable for drylands have been organized by the Australian Commonwealth Scientific and Industrial Research Organization (CSIRO) for Australian species of *Acacia, Casuarina* and *Eucalyptus* (see Boland and Turnbull, 1981); the Food and Agricultural Organization of the United Nations (FAO) for some eleven species with potential for rural improvement in drylands of nine countries in Africa, Asia and Latin America (see FAO/IBPGR, 1980); the Oxford Forestry Institute (OFI) for some 26 species from central America (see Burley *et al.*, 1986) and selected *Acacia* species from Africa (OFI, 1988).

The French government is currently supporting FAO in the initiation of exploration of species suitable for the Sahelian zone of Africa and the US government supports networks of species exploration and evaluation in Asia, including some for dry sites; both of these collaborative activities resulted from initiatives of the International Union of Forestry Research Organizations (IUFRO) in identifying problems, potential species and collaborators (see, for example, Burley and Stewart, 1985; Carlson and Shea, 1986; Iyamabo, 1987).

Such exploratory studies are supported by examination of taxonomic status through classical morphological characteristics and through determination of variation in biochemical traits; these may have value commercially (such as the gum exudates of *Acacia*, *Boswellia* and *Commiphora* species) but are used at the exploratory stage as genetic markers of variation at the population and individual levels (see, for example, Burley and Lockhart, 1985). For many species the breeding system is unknown and detailed studies are required of floral and pollination mechanisms before serious genetic improvement can be attempted (Burley *et al.*, 1986).

The original cause of population differences was natural selection, frequently leading to continuous (clinal) or discontinuous (ecotypic) patterns of variation; however, for many species and populations man's activities have modified the natural pattern and extent of genetic variation. This occurs primarily by the destruction or severe genetic impoverishment of primary gene pools through overcutting or overgrazing; it also occurs through the modification of population genetic structure in planted species (the creation of land races by either deliberate or unconscious selection). Many species are threatened by total extinction and in others individual populations are at risk even though the entire species may not be (Palmberg, 1981). There is thus extreme urgency for conservation *in situ* or *ex situ*, as discussed below.

Evaluation

The trees used in arid and semi-arid lands yield a range of productive, environmental and social benefits, as discussed earlier. Many of these are intangible and difficult to evaluate in quantitative or financial terms; characteristics such as water flow moderation, soil holding, risk insurance, counter-seasonality and generation of employment do not have clearly marketable values, and some of these traits are not easy to assess in the field. Foresters have well-defined methods of assessing production of industrial timber from plantations of relatively straight and cylindrical conifers or eucalypts, but there are as yet no standard methods of assessing trees that are multistem-med, managed by coppicing or lopping for fuel or forage, and scattered in farm or range lands (see Stewart, 1989). Livestock specialists, agronomists and soil scientists must be involved in the assessments as well as foresters.

Standard designs exist for comparative experiments to evaluate trees for plantations but modifications of these designs are urgently needed for trees to be used in farm systems such as agroforestry or in extensive pastures. The International Council for Research in Agroforestry (ICRAF) is currently investigating alternative designs (see also Burley and Riley, 1989).

Whatever the experimental design used, great advantages result from internationally coordinated trials such as those described above. The many populations and species collected during the exploratory stage can be established in experiments in a range of sites and management systems; combined analysis of resultant data yields information on genotype–environment interaction and genotypic stability that could not be obtained from the results of an individual trial.

Conservation

There is increasing global recognition of the need to conserve plant genetic resources not only for their sustainable use in current sites and for current objectives but also for exotic sites and hitherto undiscovered benefits. The emphasis is now on tropical rain forest conservation because of the implications of its loss for global climate change, but the case for conservation of representative areas and genotypes of dryland vegetation is equally strong (Palmberg, 1981).

Conservation of plant resources may be through *in situ* conservation of areas of typical vegetation, but where human population pressure is great the political and managerial difficulties of ensuring the protection and sustainable management of such areas are often insuperable. In such cases the alternative system is *ex situ* conservation whereby representative samples of seed, clonal propagules, tissue cultures or, possibly in the future, DNA libraries can be stored and propagated elsewhere, often with the intention of reintroducing material to its native habitat. For many dryland species the techniques of *ex situ* conservation and management have not yet been determined.

Genetic Utilization

A major enhancement of the productivity of arid lands may be achieved through the application of plant breeding practices to trees and shrubs as well as to pasture grasses and legumes. Because of the relatively low agricultural productivity of these lands government research and rural development agencies have devoted few resources to this in the past. Now, with increasing populations in the areas and the concomitant degradation of remaining land, the need is great.

Tree breeding follows the same principles as traditional plant breeding with selection of superior individual genotypes, their evaluation in progeny tests and their bulk propagation through special seed orchards or, in the case of vegetatively propagated plants, through clonal archives. Unlike agricultural plants, however, trees have problems associated with their size, longevity, space requirements and multiple benefits; these particularly influence experimental design and evaluation methods. Where trained staff and facilities are available multipopulation, recurrent generation breeding is desirable but, in common with all other aspects of dryland development, the lack of prior government commitment has resulted in a severe shortage of trained professional and technical staff for tree improvement.

When genetically improved material is produced it must be made available to the institutions or individuals wishing to use it on the large scale in the field. This in turn requires the development of mass propagation methods and the establishment of a government extension service able to demonstrate, monitor and advise on the implementation of new materials and systems. The propagation techniques vary with species; the traditional method for trees and shrubs is through seed but for many woody species clonal techniques are becoming widespread, often through simple, cheap rooting of cuttings; the benefits of vegetative propagation lie in the uniformity of the stock produced and the capture of total genetic improvement, including both additive and non-additive genetic effects. However, there are disadvantages among which the principal is the disease risk implicit in any monoculture.

ADOPTION IN LAND-USE SYSTEMS

There are many systems of land use that incorporate trees and shrubs; they can be most simply classified into three systems: natural vegetation, plantations and agroforestry systems.

Natural Vegetation

Historically in tropical countries rain forest has received the most attention in terms of research, management and public interest, but donor agencies and national governments are now beginning to support development and conservation of natural forest and woodland in drier zones. Apart from drastic remedies such as enforced control of human and animal population size and movement, the development and sustained use of dry vegetation, particularly woodland, requires new approaches to harvesting cycles and the introduction of new species for soil enrichment, fuel and fodder production; this type of enrichment is being applied to fallow lands after cutting existing trees and shrubs or in conjunction with planned pasture rotation.

Plantations

Plantations are deliberately planted, large blocks of trees or shrubs, generally in a monoculture. They may be undertaken at three levels for different purposes and by different institutions.

Industrial plantations   Industrial plantations are established by government agencies and by companies to support specific industries such as pulp and paper, sawmilling, furniture and occasionally fuel through charcoal (steel manufacture) or firewood (tobacco drying). These are rarely established in drylands because the rates of tree growth are slow and the financial and economic rates of return are low.

Community woodlots Community woodlots are established by rural communities on communally owned land and were the basis of the movement towards social and community forestry in the 1970s and early 1980s. Under this scheme plantations were intended to produce primarily firewood and light building poles and possibly fodder from the trees themselves or from associated ground vegetation that returned with the onset of protective measures that reduced cattle grazing, uncontrolled cutting and fire.

For communities that had no experience of tree planting and maintenance, government forestry and rural development departments often provided information, advice and, in some cases, trees. They entered into various types of cost-sharing with the community leaders but global experience has shown that greatest gains occur when the communities themselves undertake all the stages of woodlot development.

The products from such woodlots were either intended to be distributed to members of the community or were sold and the proceeds devoted to community works such as building of schools or roads. In practice nepotism or poor management led to inequalities in the distribution of benefits and many governments and donor agencies began to support projects that encouraged farm woodlots.

The concept of a plantation being established specifically for the domestic requirements of a large community is not new; early in the twentieth century fuelwood plantations were established around many towns in Africa, e.g. Ibadan, and indeed many were established between towns as fuelwood sources for railway engines in eastern, central and southern Africa. The Sahelian droughts of the late 1970s precipitated considerable efforts to plant trees for fuelwood and to halt desertification; however, these were unsuccessful largely because the species chosen were not optimal, the costs of establishing them were far in excess of any likely financial return with such slow growth and the local social structure and land tenure did not encourage the maintenance of trees. Also in most cases the tree planting activities were the responsibility of the government forest service, which itself had little experience of forestry in difficult dry conditions and did not involve cooperation from other development ministries and agencies.

Farm woodlots In drylands farmers have not traditionally planted trees other than for fruit production. Trees have always been considered an inexhaustible natural resource that does not need human intervention. Now, in common with many other common property resources, trees and woody shrubs are recognized as declining in availability and requiring positive intervention. In more humid lands farmers have planted farm woodlots successfully for subsistence and marketing without compromising agricultural productivity or by substituting purchased food for home-grown crops.

It is difficult to encourage the planting and maintenance of trees in drylands, particularly where pastoralism is prevalent, and it requires the careful choice of species and system, and the provision of reputable supplies of seed or other planting material; this in turn requires the creation of a government extension service that is sensitive to people's perceptions of need and possibly the strengthening of local transport and marketing infrastructures.

Many donor agencies and individual governments now recognize that, provided land tenure can be assured for the farmer, projects that support individual tree planting are more likely to succeed than those requiring communal effort. The benefits of farm woodlots to the individual land owner include traditional forest and tree products, contributions to environmental sustainability and cash income.

Agroforestry Systems

Agroforestry is a collective name for land-use systems and technologies where woody perennials (trees, shrubs, palms, bamboos, etc.) are deliberately used on the same land management unit as agricultural crops and/or animals, either in some form of spatial arrangement or temporal sequence. In agroforestry systems there are both ecological and economic interactions between the different components (see Lundgren, 1980). Economic here includes the social and environmental benefits described earlier.

There is in fact a continuous spectrum of land use from pure crop or livestock production to pure forestry with all combinations between them (see Wood and Burley, 1989) (see Figure 7.1).

There are four main groups of true agroforestry systems that involve planting trees and shrubs, some of which are less suitable for drylands than others.

Zonal systems This group includes systems in which lines of trees are planted systematically in relation to the agricultural crop. The lines may be single rows or multiple rows of trees and the spacing between lines within which cropping is possible varies. The most widely discussed system is currently alley cropping (hedgerow intercropping) in which the tree compon-

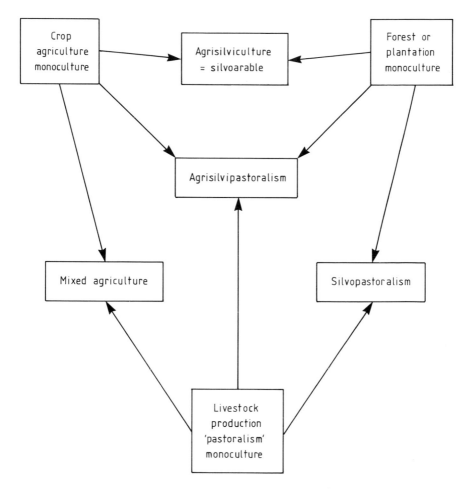

Figure 7.1 'Agroforestry' in the continuum of land husbandry systems

ent may yield fuel, poles, fodder or, frequently, mulch to reduce the inorganic fertilizer input; a common combination is maize with *Leucaena leucocephala*, ahough the tree species is not well adapted to acid soils or very dry conditions and it is also susceptible to attack by a psyllid insect. Other species including *Gliricidia* and *Sesbania* have promise for these conditions but the system is less suitable as conditions become drier and overall crop productivity declines.

Linear systems   Alley cropping may itself be arranged linearly or with the lines of trees along contours, particularly where soil conservation is a primary

Figure 7.2 *Casuarina* planted on farm border in Egypt as a windbreak and for the production of fuelwood and fodder. (Photo J. Burley)

objective. However, linear arrangements usually refer to shelterbelts and windbreaks in which single rows or belts of trees are separated by considerable distances. The objectives of these are to provide micrometeorological improvement of the environment for crops or homesteads respectively, especially the reduction of wind damage and evaporation. They are commonly established at higher latitudes, e.g. Sahelian and Sudanian zones in Africa, and can yield many products including food, fodder and fuel while reducing environmental stress, soil erosion and cash shortages, and enhancing crop production (e.g. Catterson *et al.*, 1987).

Trees are commonly planted on farm boundaries for demarcation and production, contour strips to aid soil holding, as living field fence posts and as hedges for fodder banks or cattle enclosures or exclosures (see Figure 7.2).

Mixed systems   All of the above systems include mixtures of trees or shrubs and agricultural crops. However, the term mixed system is commonly reserved for those land uses in which trees are left standing over, or planted deliberately in, agricultural fields or pasture. The species are commonly nitrogen-fixing and are well-known to local populations as sources of multiple benefits. They can contribute to local economics and environments in the same way as farm woodlots or linear plantings but at a less intense scale; they therefore occupy a smaller proportion of productive land and in some cases actually enhance crop production through mulching or shading effects, e.g. *Acacia albida* enhances the productivity of maize, sorghum and millet because it is able to produce its own leaves in the dry season when there is no demand from the crop and its fallen leaves and root breakdown products increase soil fertility (see Felker, 1978; Catterson *et al.*, 1987).

Where pastoralism is predominant the protection of newly planted trees and shrubs is difficult and often expensive, particularly when fencing materials are not locally available. However, trees may be used to enrich pastures directly and increase shade for animals, while they provide an excellent source of fodder where cut-and-carry systems for stall-fed cattle are accepted (see Figure 7.3).

Sequential systems   This group includes systems of land use in which trees and crops are raised on the same unit of land at different or overlapping times within seasons, between seasons or between years. In a sense traditional shifting cultivation is an agroforestry system since existing natural vegetation is felled and burned in order to gain the nutrients for one or two years of crop growth; these are then followed by a period of fallow in which natural

Figure 7.3   Camels grazing on *Acacia tortilis* in Baringo District, Kenya. (Photo C. Fagg)

vegetation regenerates itself and soil nutrients. When human population pressures are not excessive and the rotation period is long (20–30 years) this system is appropriate and sustainable; however, throughout the tropics, as pressures have increased and cycles have shortened, fallow periods have not been able tò regenerate the site and degradation and desertification have ensued.

Taungya is a centuries-old technique for the establishment of tree plantations at the same time as land is used for agricultural cropping. The system employs licensed or directly employed cultivators to plant the trees and keep them weed-free until they mature sufficiently to survive weed competition; until this time the cultivator raises his own or governmental crops. Afterwards he moves to another plot.

More recently attention has been paid to the development of sequential systems in which more intensive rotations and more intimate mixtures of tree and crop are managed on small farmholdings. ICRAF is currently designing experiments to study the effects of such mixtures and different methods of tree management (coppicing, lopping and pruning) on productivity and sustainability.

Such sequential systems may yield a range of productive and environmental benefits while also addressing the common social problem of the lack of land tenure by individuals.

RESEARCH NEEDS FOR PLANTS IN DRYLAND DEVELOPMENT

Natural Vegetation

Throughout the arid and semi-arid regions little research has been done on the management of natural vegetation. Trees, shrubs and herbs have been used traditionally but virtually nothing is known about the effects of different animal stocking densities on the survival and growth of individual plants, species or ecosystems; the carrying capacity of the land is not well quantified although it is recognized that in many areas current stocking grossly exceeds the sustainable capacity. Research is therefore needed on the regeneration of indigenous vegetation and on socially acceptable means of controlling pastoralism and stocking density (see, for example, UNESCO, 1978; Maydell, 1986).

Choice of Species and Population for Planting

As shown earlier, one of the major contributions to dryland productivity can be the genetic improvement of trees and shrubs for a range of plantation systems. The first step in the improvement process is the selection of the optimum species and population, and although some research is in progress, as described above, there is an urgent need to expand this to other countries and species.

New research is needed on the determination of internationally comparable methods of assessment of trees and shrubs. The standard forestry measurements for industrial plantation species, including height, diameter, taper and straightness, are not relevant to species required by farmers for fuelwood, fodder or fencing (see Stewart, 1988); the shape, size and properties of trees required in rural forestry are different and for some end uses a totally different range of characteristics and analyses will be required (see, for example, Oduol *et al.*, 1986, who assessed the genetic variation within *Prosopis* for the sugar and protein contents of pods for fodder).

The benefits of the international programmes can be enhanced and made more widely available by network research and dissemination of results. However, for the results to be implemented, there is an equally pressing need for the development of a professional forestry or agroforestry service in many countries and the enhancement of extension and training services.

Propagation and Planting Techniques

Since so little tree and shrub planting has been done in drylands there is ample room for improvement of techniques. Traditionally woody plants are propagated from seed but vegetative propagation is practised by cuttings, by grafting for some fruit and nut species, and by tissue culture and micro-propagation (see several papers in Felker, 1986).

The traditional industrial nursery system raised plants some 10–30 cm tall in bags or tubes containing up to 1 kg of potting mixture or local soil; recently the trend has been to develop systems with small, multiplant container trays in which the soil content per plant is 0.1 kg or less, thus reducing transport and handling costs. However, in dry zones these smaller plants may have a reduced chance of survival and research is needed into the optimum propagation method, container size, plant pre-treatment before delivery to the field and the feasibility of including water-retaining polymers in soil mixtures.

Ground preparation before planting and postplanting culture are also important determinants of the survival and growth of woody plants, but again little research has been done to date on optimum techniques for drylands. Three major sources of existing information are Ghosh (1977), Goor and Barney (1968) and Kaul (1970).

Manipulation of Soil Microbial Associates

Many trees and shrubs naturally form associations with root-inhabiting micro-organisms. These include the rhizobial bacteria that form nodules on the roots of many leguminous species; they fix nitrogen from the atmosphere that becomes available to the host plant in varying amounts (still to be determined for most conditions). Some tree species that are not legumes form associations that fix nitrogen with actinomycetes (primitive fungi). Most tree and shrub species are believed to form root associations with mycorrhizal fungi that assist in the uptake of soil nutrients, particularly phosphorus. Many leguminous trees have tripartite symbioses with rhizobia and vesicular–arbuscular mycorrhizae (see, for example, Roskoski *et al.*, 1986).

The extent, nature and effects of these associations requires research. While the trees and shrubs show genetic variation between populations within species it may also be possible that indigenous or introduced microorganisms also vary genetically in a systematic pattern. Methods are needed for the culture, transfer, inoculation and assessment of effects of these associates. The international microbial research centres (MIRCENS) maintain large numbers of rhizobial cultures that can be made available for field trials.

Determination of Human Perceptions

A large proportion of recent development initiatives, including the rapid international spread of enthusiasm for agroforestry and dry zone forestry, has been promoted by scientists and professional land managers or development agencies, often from outside the particular region or country of concern. They have frequently overlooked or minimized attention to the perceptions of the rural populations that they intend to help. There is an urgent need for site-specific sociological and socioeconomic research to determine the views of local individuals and communities on their resource needs and the acceptable ways in which they could be met. The perceived wisdom of the scientist is not necessarily relevant to the culture and organization of the indigenous people.

A CASE STUDY—THE OFI PROGRAMME FOR ARID ZONE HARDWOODS

The Oxford Forestry Institute (OFI, originally the Imperial and until recently the Commonwealth Forestry Institute) has for some 25 years undertaken studies of the genetic resources of tropical trees, following the various stages outlined above. These concentrated initially on industrial pine species suitable for paper and saw-timber, and advanced breeding populations of some species are beginning to emerge as the basis of national breeding programmes in many countries. In the late 1970s and early 1980s attention spread to hardwoods for industrial plantations intended to produce fine timbers for furniture and major construction.

At the start of the 1980s many donor agencies, national governments and non-governmental organizations began to support land-use projects that involved and benefitted rural populations, particularly those dependent on land that was dry or degraded. The OFI therefore initiated similar studies of species suitable for such conditions. With one exception, all the species chosen in these 25 years of research have been indigenous to central America where the OFI has built strong links with local organizations and where it has contributed significantly to the knowledge of the local flora. However, the main benefit has been that the species selected have great potential for planting in appropriate site types worldwide. The one exception to the central American origin is a study of indigenous African *Acacia* species that are promising for dryland development.

In 1982 the OFI started the study and genetic improvement of some 27 central American species, mainly legumes, that are well known in their native countries as being suitable for fuel, fodder, poles or soil improvement while tolerating extreme environments. The region is rich in such species but many wild populations of several species are threatened with extinction or severe genetic impoverishment. The programme therefore included the exploration of the natural range of each species, sampling for taxonomic confirmation, and collection of seed from selected populations for standardized, replicated field trials throughout the semi-arid tropics.

In 1984 an international trial was initiated in which seed of the 27 species was sent to 63 countries for over 150 separate trials (see Burley *et al.*, 1986; Hughes and Styles, 1984; and Stewart, 1989). Preliminary results indicated the widespread potential of one species, *Gliricidia sepium*, at least at the more humid sites within semi-arid lands; for this species range-wide provenance collections have been made from 28 populations in seven countries. (Some collections are made from identified individual trees so that half-sib progeny tests can also be established to estimate population genetic parameters.) Since early 1987 samples of seed have been distributed for 176 trials in 51 countries, again with suggested standard designs (see Stewart, 1988, 1989). Preliminary results from some of these experiments demonstrate considerable variation between populations in survival, form and productivity. They have also indicated the difficulties of assessing trees in the managerial system, tree shape and size that rural populations require.

The advantages of such international collaborative research lie in the well-documented collections of test material, the standardization of design and some assessment, and the ability to estimate genotype–environment interaction effects and hence the stability or plasticity of each population. This allows the identification of optimum sources for given site types, even when an individual site has not been tested, and the development of breeding strategies and populations for the future.

This case study includes only the first genetic steps in the introduction, improvement and adoption of a species in land use. Although the experiments provide for testing the tree species in a range of cultural systems, there remain the major problems of developing large-scale methods for the production of optimum stock, determining on-farm and on-range productivities, creating an appropriate extension service and identifying farmers' perceptions before widespread adoption can be expected. It must be hoped that the increasing political awareness of the needs of dryland inhabitants will bring the additional resources needed to address these issues, and sustain and enhance the productivity of the arid and semi-arid lands throughout the tropics and sub-tropics.

REFERENCES

Boland, D. J. and Turnbull, J. W. (1981). The selection of Australian trees other than eucalypts for trials as firewood species in developing countries, *Australian Forester*, **44**, 235–46.
Burley, J. and Lockhart, L. A. (1985). Chemical extractives and exudates from trees, in *Plant Products and the New Technology* (Eds. K. W. Fuller and J. R. Gallon), pp. 91–102, Clarendon Press, Oxford, England.
Burley, J. and Riley, J. (1989). Design criteria and evaluation techniques in agrofor-

estry. Invited Paper, Centennial Symposium, Department of Forestry and Natural Resources, Edinburgh University, Scotland, July 1989.

Burley, J. and Stewart, J. L. (Eds.) (1985). Increasing productivity of multipurpose species, *Proc. IUFRO Planning Workshop for Asia on Forest Research and Technology Transfer*, International Union of Forestry Research Organizations, Vienna, Austria, 560 pp.

Burley, J., Hughes, C. E. and Styles, B. T. (1986). Genetic systems of tree species for arid and semiarid lands, *Forest Ecology and Management*, **16**, 317–44.

Carlson, L. W. and Shea, K. R. (1986). Increasing productivity of multipurpose lands, *Proc. IUFRO Research Planning Workshop for Africa Sahelian and North Sudanian Zones*, Canadian Forestry Service, Hull, Canada, 333 pp.

Catterson, T. M., Gulick, F. A. and Resch, T. (1987). Rethinking forestry strategy in Africa: experience drawn from USAID activities, *Desertification Control Bulletin*, **14**, 31–7.

FAO/IBPGR (1980). *Genetic Resources of Tree Species in Arid and Semi-arid Lands— A Survey for the Improvement of Rural Living in Latin America, Africa, India and South West Asia*, FAO, Rome, Italy, 118 pp.

Felker, P. (1978). State of the art: *Acacia albida*, USAID Grant Report Afr-C-1361, University of California, Riverside, Calif., USA, 87 pp.

Felker, P. (Ed.) (1986). *Tree Plantings in Semi-arid Regions*, Elsevier, Amsterdam, Netherlands, 444 pp.

Ghosh, R. C. (1977). *Handbook on Afforestation Techniques*, Forest Research Institute and Colleges, Dehra Dun, India, 411 pp.

Goor, A. Y. and Barney, C. W. (1968). *Forest Tree Planting in Arid Zones*, Ronald Press, New York, 409 pp.

Hughes, C. E. and Styles, B. T. (1984). Exploration and seed collection of multipurpose dry zone trees in Central America, *International Tree Crops Journal*, **3**(1), 1–31.

Iyamabo, D. E. (Ed.) (1987). *Tree Improvement and Silvo-pastoral Management in Sahelian and North Sudanian Africa*, International Union of Forestry Research Organizations, Vienna, Austria, 196 pp.

Kaul, R. N. (Ed.) (1970). *Afforestation in Arid Zones*, Junk, The Hague, Netherlands, 435 pp.

Lundgren, B. (1980). The use of agroforestry to improve the productivity of converted tropical land, Report prepared for Office of Technology Assessment, US Congress, IUFRO, Nairobi, Kenya, 82 pp.

Maydell, H.-J. von (1986). *Trees and Shrubs of the Sahel; Their Characteristics and Uses*, Deutsche Gesellschaft für Technische Zusammenarbeit, Eschborn, Germany, 525 pp.

OFI (1988). *Sixty-third Annual Report, 1987*, Oxford Forestry Institute, Oxford, England, 56 pp.

Oduol, P. A., Felker, P., McKinley, C. R. and Meier, C. E. (1986). Variation among selected *Prosopis* families for pod sugar and pod protein contents. In *Tree Plantings in Semi-Arid Regions* (Ed. P. Felker), Elsevier, Amsterdam, The Netherlands, 423–431.

Palmberg, C. (1981). A vital fuelwood genepool is in danger, *Unasylva*, **33**(1), 22–30.

Roskoski, J. P., Pepper, I. and Pardo, E. (1986). Inoculation of leguminous trees with rhizobia and VA mycorrhizal fungi. In *Tree Plantings in Semi Arid Regions* (Ed. P. Felker), Elsevier, Amsterdam, The Netherlands, 57–68.

Stewart, J. L. (1988). *International trial of Central American dry zone hardwood species. Field manual*. Oxford Forestry Institute, Oxford, England, 81 pp.

Stewart, J. L. (1989). Genetic improvement of Central American dry zone multi-purpose species; a case study, in *Nursery Technology for Afforestation of Arid and Semi-arid Regions* (Eds. S. Puri and P. K. Khosla).

UNESCO (1978). Tropical forest ecosystems. A state of knowledge report, UNESCO/UNEP/FAO, Natural Resources Research 14, 683 pp.

UNESCO (1979). Map of the world distribution of arid regions; explanatory note, UNESCO, Paris, France, 54 pp.

Wood, P. J. and Burley, J. (1989). The potential of agroforestry. Background Paper 2 in *Agroforestry: potential, current UK expertise, and research needs—a guide to ODA strategy* (Eds. J. Burley and N. Wilson), Oxford Forestry Institute, Oxford, England, 93 pp.

CHAPTER 8

The Conservation and Management of Semi-arid Rangelands and Their Animal Resources

M. COE

INTRODUCTION

The ecologist studies the interrelationships between living things and their environment. Such interrelationships are incredibly complex, for detailed studies over the last 60 years have revealed that not only is there a large degree of interdependence between living things but their influence on their non-living surroundings is also of profound importance. The richness of the flora and fauna of any one area has a strong correlation with its latitude and climate, so that tropical ecosystems, which lie almost entirely in the developing world, are under intense pressure, which could in the next hundred years lead to the disappearance from the face of the earth of thousands or even hundreds of thousands of species (Myers, 1979).

In the last 25 years the term 'ecology' has been greatly devalued as it has been increasingly employed (often incorrectly and inappropriately) in a political context. Colinvaux (1980) recently summarized this problem when he wrote, 'Ecology is not the science of pollution, nor is it environmental science. Still less is it the science of doom.' This does not of course mean that all ecologists are engaged in esoteric science that has no relevance to the predicament of modern mankind, but rather that we have a great deal to learn from these scientific natural historians, who have spent the last 60 years trying to work out how the natural world functions. Such studies are not easy, for even the most seemingly simple ecosystem often turns out on detailed examination to be amazingly complex. Contemporary environmental

Techniques for Desert Reclamation
Edited by A. S. Goudie
© 1990 John Wiley & Sons Ltd.

concerns such as pollution are problems created by humans in their determination to exploit the animate and inanimate resources of the biosphere,
largely for the benefit of the developed world. Ecologists can tell us what the
observed effect of these agents is on the natural world, but the removal or
correction of these effects lies with industrialists and politicians, not with the
ecologist. Herein lies the dilemma of both the developed and the developing
world, for the time perspectives of the ecologist are of an entirely different
scale. The ecologist may be able to visit an area to provide a rapid Environmental Impact Statement in the face of impending development, but natural
systems develop at differing rates depending on local conditions, and only
long-term monitoring over many years will provide the degree of detail
required to understand, and more importantly predict, the direction of
change in the face of a particular form of intervention. By contrast political
regimes and well-meaning development plans are constrained by much
shorter timescales of perhaps 5 or 10 years, even in one-party States, which
are by no means uncommon in modern Africa (Coe, 1986).

When we recognize that 1200 million people live on the world's environmentally sensitive savannas, with 65% of this number in Asia, 18% in Latin
America and 17% in Africa (Harris, 1980), it is apparent that we must
explore the common ground between the ecologist and the development
planner if we are to solve their ever-present problems of survival in the harsh
unpredictable environments that they occupy. To close the gap between the
wealthy nations in the north and the abject poverty that enslaves so many of
the world's population in the arid south, requires a pooling of the knowledge
accumulated by these field biologists and the seemingly more practical objectives of those involved in planning development programmes.

In recent years the world has become increasingly aware of the problems of
the Sahel Zone of Africa, due to extensive television coverage of this tragedy
in the West and the amazing efforts of people like Bob Geldof who made it
his personal mission to prevent such a disaster ever happening again. Yet
although we all willingly provide money to wipe the hideous image of starving
children from our television screens, such famine aid does little more than
ensure the survival of these people until the next time their parched lands are
struck by drought, when the effects, in the absence of a dramatic decrease
in their numbers or the means to feed their population from their own
resources, will produce a disaster of even greater proportions.

VEGETATION AND CLIMATE IN THE ARID WORLD

The world's arid regions are commonly classified according to their annual
rainfall, which may lie below 300 mm in semi-arid regions and between 500
and 1000 mm in semi-humid savanna (Harris, 1980). In Africa, Dudley-Stamp
and Morgan (1973) considered that bushland environments received

300–700 mm, woodland 800–1200 mm and wooded grassland 700–1800 mm. In terms of the utilization and management of these areas it is frequently the length of the dry season that acts as an important limiting factor in the utilization of arid lands, for as Harris (1980) has pointed out, the dry season may last for 2.5–7.5 months in the Intertropical Zone and 7.5–10.0 months in the Tropical Zone. The biggest climatic problem in these areas, however, is the large annual variation in rainfall, which we may expect to vary from year to year by 10–20% of the norm in humid savanna, 20–25% in sub-humid savanna and 25–40% in semi arid savanna.

In areas of eastern Kenya rain falls in a bimodal pattern, during the so-called 'long rains' (late September–October to early January) and the 'short rains' (mid March to mid May). There is, however, little difference between the mean annual totals for these two seasons, although Tyrrell and Coe (1974) demonstrated that, using a series of rain gauges in the Tsavo National Park, the short rains were more predictable (less likely to fail) than the long rains. Additionally, it appeared that many local plant species flower extensively in relation to this greater predictability of the short (or second) rains, and not at all or only sparsely in the long (or second) rains, a phenological feature that may well be related to the accumulation of a larger quantity of dead organic matter in the long dry season (May to October) compared with the much shorter dry season between January and March. This greater build-up of organic matter will potentially release a much larger pulse of nutrients for plant growth and reproduction compared with their much smaller availability in March. The great year-to-year variation in mean annual precipitation is well illustrated in the Voi area of Kenya (03°24'S, 38°34'E), which lies adjacent to the Tsavo East National Park whose 64 year mean in 1974 was 538 mm, but the range over this period extends from 184 to 1201 mm (Anderson and Coe, 1974). In addition to this large annual variation there is also a tremendous local variation, especially where sufficient rain gauges are deployed to pick up such patterns of microdistribution. Records for the Tsavo National Park recorded by the Tsavo Research Project demonstrated that the range over the whole Park area of 21 000 km^2 lay between 209.15 and 1073.42 mm in 1968 and 168.20 and 693.75 mm in 1973 (Glover, 1970, 1973).

A further factor of great importance in arid regions is the observation that as the degree of aridity increases, there is a stronger tendency for rain to occur in short sharp showers, which severely limits its effectiveness for plant growth, since the precipitation is likely to exceed the infiltration capacity of the soil, particularly when these possess a high clay fraction, a condition by no means uncommon on basement-derived soils (Harris, 1980). We may observe the importance of this potential limit to primary production if we once again examine the rainfall data for Voi (Kenya), where between 10 and 16% of the annual total may fall in a single 24 hour period, while if we take the highest single day's rain in each month, 40% (1970) may fall in these 12 days (Anon, 1978). Hence we might expect any predictive relationship between mean annual rainfall and primary production to become less reliable at higher levels of aridity.

On vegetated ground, water is lost by evaporation from the ground surface and from plants by transpiration from their foliage. Together, these pathways of water loss are referred to as evapotranspiration, but its measurement in practice is difficult and complicated. Most ecologists who are interested in the magnitude of primary production in a given rainfall regime are concerned with the potential evapotranspiration (PET) or the water loss from a soil/vegetation complex with adequate moisture and the actual evapotranspiration (AE), the water loss from a soil/vegetation complex under natural conditions. The estimation of the former is possible but the measurement of the latter extremely difficult since it depends on the ground cover of the vegetation, its height, density and the depth of the rooting system (Griffiths, 1976). In spite of these difficulties, however, a number of systems have been developed for the calculation of evapotranspiration from meteorological data—especially temperature.

Walter (1954) and Whittaker (1970) have established a predictive relationship between rainfall and primary production, while Rosenzweig (1968) described a similar relationship between net above-ground annual primary production (NAAP) and AE. The last-named author's equation for calculating NAAP was

$$\log_{10}\text{NAAP} = \log_{10}\text{AE}(1.66 \pm 0.27) - (1.66 \pm 0.07)$$

where NAAP = net above-ground annual primary production (g m^{-2} a^{-1})
AE = annual actual evapotranspiration (mm a^{-1}).

Here, we may consider AE to be a simultaneous measurement of water availability and solar radiation, since both may limit the rate of photosynthesis (*sensu* Rosenzweig, 1968). In both arid and semi-arid ecosystems PET may normally be expected to be greater than annual precipitation (AP). Coe *et al.* (1976), in a study of the biomass and production of large African herbivores in relation to rainfall and primary production, suggested that AE could be considered equal to AP up to 700 mm a^{-1}.

Few measurements of the actual relationship between NAAP and rainfall are available for African savannas but Phillipson (1975), estimating NAAP from precipitation in the Tsavo East National Park in eastern Kenya, showed that his figure of 700 g m^{-2} a^{-1} was close to that obtained from direct measurements at Buchuma on the Park's eastern boundary, namely 648 g m^{-2} a^{-1} (Cassidy, 1973). This last named author showed that near maximum yields were obtained within 50–100 days of rain, when they averaged 32–68 kg ha^{-1} d^{-1}. The more detailed studies of McNaughton

(1985), working on the grasslands and large herbivorous grazers of the Serengeti Plains in Tanzania and the contiguous Mara Game Reserve in Kenya, has demonstrated that control (exclosure) productivity is linearly related to rainfall ($r = 0.695$, d.f. $= 18$, $p < 0.001$), averaging 357 g m^{-2} a^{-1}, and 1.89 g m^{-2} d^{-1} in the growing season. These estimates, however, referred to the control plots and the relationship was considerably reduced when other more physiographically and climatically diverse sites were included ($r = 0.416$, $p < 0.05$, d.f. $= 26$). Actual productivity, in the presence of large herbivores, showed substantially greater productivity than the control sites, averaging 664 g m^{-2} a^{-1}. A similar feature was evident with growth rate, which on the natural sites reached 3.84 g m^{-2} d^{-1}. It is, however, interesting to note that actual production showed an even closer correlation with grazing intensity ($r = 0.619$, $p < 0.001$, d.f. $= 26$) than it did to rainfall, an important indicator of the powerful effect that a diverse herbivore community of grazers have on grassland productivity—a feature that may be thought of as the 'lawnmower effect'. Indeed, as McNaughton (1985) points out, 'The ecology of neither the plants nor the animals in the Serengeti ecosystem can be understood in isolation; many traits of both suggest coevolution among trophic web members.' Thus is seems reasonable to assume that an understanding of semi-arid and arid ecosystems may lie in assessing the role of all the animal and plant components of the system, rather than individual entities or the one-way movement of material, for if we are to manage and utilize them, the removal or replacement of even some of the less numerous or obvious components may lead to their collapse or at least a severe reduction in their natural levels of secondary production.

Clearly our understanding of plant production is a vital prerequisite to estimating the carrying capacity of these ecosystems, whether this is from the point of view of pastoralists and their domestic stock, wildlife or a combination of both. Since we have already observed that there is a close relationship between rainfall and primary production it is clear that these highly variable climatic regimes must exhibit great interannual variations in annual primary production. Phillipson (1975) used annual and monthly rainfall records (1969–72) for the southern section of the Tsavo (East) National Park to estimate variations in NAAP and showed that both before and during a period of intense drought, primary production ranged from 545 g m^{-2} a^{-1} in 1969 to 210 g m^{-2} a^{-1} in 1971. Using these data he was able to demonstrate that the catastrophic mortality of elephants during the 1970–1 drought coincided with the areas whose predicted primary production was <200 g dry wt m^{-2} a^{-1}, with the centre of the most intense mortality lying in those areas producing <100 g dry wt m^{-2} a^{-1}. These observations showed a closer correlation with the centres of mortality than the 10 inch 10% rainfall probability isohyet used by Corfield (1973). Since most of this mortality occurred in the vicinity of permanent water it provided strong circumstantial evidence that starvation was the main cause of death rather than lack of water and/or desiccation (Phillipson, 1975).

If we examine a semi-arid area like the Tsavo National Park it is possible not only to predict the amount of primary production potentially available to the large herbivores but also to detect those periods in which available forage will be inadequate to sustain the existing populations comprising the wild herbivore community of the National Park. Phillipson (1975), in an elegant attempt to predict the effect on elephants of such climatic variation, estimated that in non-drought years they would consume no more than 6.4% of the primary production, while in drought years this would rise to 8–12%. Using faecal output data (Coe, 1972) he estimated that in 1970 this figure would have been 9%. Using estimated mean rainfall figures derived from 20 rain gauges in Tsavo East and faecal pellet data for the whole range of large herbivores (Coe and Carr, 1983; Coe, 1983), it is possible to estimate that these mammals consume 106.31 kg dry wt $km^{-2} d^{-1}$ in the dry season and 131.91 kg dry wt $km^{-2} d^{-1}$ in the wet season (Coe and Kingston, 1988). This consumption suggests that elephant forage contributes 83.8% and 69.9% of the total in wet and dry seasons respectively. It has been proposed that when elephant consumption rises to nearly 21% of NAAP in a severe drought, this will be responsible for drastic mortality (Phillipson, 1975). Coe and Kingston (1988) suggest that the large herbivore community will remove up to 10.7% of NAAP in an average year. It is here worth noting that, depending on the gross assimilation efficiency of wild herbivores, up to 25% of the food consumed by ruminants, 50% of that consumed by zebra and nearly 80% of that taken by elephants will be returned to the soil surface as dung, where it may be potentially recycled for further plant growth. Its rate of removal and reincorporation, however, will also depend on the season, for coprophagous dung beetles are maximally active in the wet season when the dung will be rapidly recycled, while in the dry season termites are the main agencies of removal (Coe, 1977), which will immobilize the nutrients for some time before they are released, mainly through the agency of termite predation. Additionally we may calculate that termites are also responsible for elevating up to 3.95 kg of soil per kg elephant dung (fresh weight), which incorporates organic matter into the soil, enhances soil aeration to a depth of up to 0.5 m and increases water penetration in soils compacted by rain and trampling. In the event that such wildlife populations are removed or depressed, this important agency of soil conditioning and nutrient cycling will be lost. In pastoral areas, where the species-rich wild fauna is replaced by a narrower spectrum of domestic species, the consequent decrease in the diversity of dung beetles, many of which are species- and time-specific in their activity, will reduce the efficiency of the decomposition of faecal material. Recent work on the use of the antihelminthic drug Invermectin on cattle (Wall and

Strong, 1987; Coe, 1987) has suggested that their careless use could lead to a complete collapse of the dung-incorporating fauna in these arid regions.

Thus we observe that in relatively undisturbed systems, dominated by very diverse communities of wild herbivores, vertebrate–invertebrate interactions and the vegetation could be vital elements in their efficient functioning, which may explain why the productivity of some of these ecosystems appears to be depressed in areas where wild mammals are either reduced or displaced.

WILDLIFE AND THE ARID LANDS

Measurements and estimates of NAAP tell us the amount of plant material being produced each year under the influence of a given climatic regime. It does not, however, tell us how much of this production is potentially and actually available as food for herbivores. To a large degree this is dependent on the growth form of the vegetation, with grasslands dominated by mono-cotyledons turning over about 30% of their biomass per annum and woodland/forest ecosystems turning over between 4 and 6% of their biomass (expressed as the production/biomass ratio) (Coe, 1980a). Clearly in extremely arid environments dominated by annual grasses the percentage of the biomass being turned over will be very much higher, while in semi-arid bushland and scattered tree grassland, with a mixture of shrubs, trees and grass, the *P/B* ratio will fall between 0.04 and 0.30. Since these ratios are dependent on the growth form of the vegetation and their life history strategies, irrespective of the climatic zone in which they occur, it is possible to estimate primary production from measurements of plant standing-crop biomass (Peters, 1980).

The wildlife communities of the African savannas have been studied extensively over the last 25 years, during which time it has been demonstrated that these diverse assemblages of large herbivores and their predators have evolved sophisticated means of avoiding or greatly reducing competition. Lamprey (1963) was the first wildlife biologist to examine the means whereby 14 common ungulate species (living in the Tarangire Game Reserve in Tanzania) could coexist. He concluded that they did so by occupying different habitats, selecting different food types, through marked seasonal movement, utilizing different feeding levels and dispersing to different dry season refuges. To actually prove that species are competing in the field is extremely difficult, but Sinclair (1974), in a study of buffalo on the Serengeti Plains, showed that during the dry season a small proportion of the huge white bearded wildebeest population enter the woodlands, where their grazing and trampling greatly reduce the forage available to buffalo, and in consequence the condition of many buffalo is greatly reduced. This competitive effect resulted in the reduced fecundity of the buffalo and a consequent levelling off of their numbers in 1970–3, although the wildebeest numbers continued to rise. These observations are of the greatest importance, for they demonstrate that it is not necessary for the whole population of a particular ungulate species to overlap with a potential competitor. Even the temporary overlap of a small proportion of the population can have a profound effect on the dynamics of another species. Similarly, a study by Field (1972) in the Ruwenzori National Park, Uganda, showed that the grass species taken by six herbivore species differed not only in relation to the species present but also their abundance (availability). Thus although there can be a considerable preference overlap between species, specialization of herbivore feeding habits tends to greatly reduce competitive interactions, even though the dry season represented a period of food shortage for a number of species. In other words, during periods of abundance, overlap will place few pressures on the herbivores, but during the extended dry seasons in arid environments there are likely to be severe limitations on animal numbers and the biomass sustained, especially where natural movements are restricted through settlement and other forms of direct or indirect development.

The strategy that a large herbivorous mammal adopts will ultimately depend on the success of its reproduction and the number of its offspring that survive to the next generation. A number of authors have recently shown that one of the most important of these life history phenomena is that of body size, which may be allometrically scaled to gestation time, growth rate, intrinsic rates of natural increase, birth rate, etc. (Blueweiss *et al.*, 1978; Western, 1979). These relationships are especially important when we consider the manner in which a particular mammal species or a group of species utilize their habitat. Although it is common to equate increase in body size with factors such as predator avoidance, if they live in an environment with an extreme and variable climate, such an increase will confer advantages in respect of the species ability to extend its 'foraging radius' (Pennycuick, 1979). This hypothesis is of special significance for migratory species in seasonal grassland environments, where the large herbivores can increase their food intake, which easily compensates for the energy expended in adopting a migratory strategy.

We have already observed that seasonal and long-term variations in rainfall have a profound effect on the plant production available to herbivores. Regions with short wet seasons (either annually or periodically) result in sudden surges of short-term food availability. Such periodicity would be expected to favour short generation time organisms such as insects, small mammals and some birds which can increase their numbers rapidly. The larger mammals, however, may increase their condition in such a season, enhance the survival of their young or even increase their numbers slightly, but they are not capable of taking the same advantage of these conditions as smaller organisms. The periodic drought that struck the Tsavo National Park in 1970–1 and resulted in severe elephant mortality had the same effect in 1960–1 when over 50% of the wildebeest population of the Nairobi National Park died of starvation (Foster and Coe, 1968). Thus during periods of drought it seems likely that herbivores will have to range over larger areas in order to satisfy their food requirements. This contention is supported by the observations of Leuthold and Sale (1973) and Leuthold (1977), who radio-tracked elephants in Tsavo East, and provide data on a series of radio-revealed home ranges of animals in Tsavo West and then northwards into the more arid regions of Tsavo East. I have taken their home ranges and plotted rainfall isohyets on them using Fourier-derived patterns generated by Cobb (1976), from which it is possible to estimate the mean rainfall received within individual home ranges. These data reveal that a significant negative linear correlation exists between mean rainfall and home range size (combined male and female $r = -0.542$, d.f. $= 26$, $p < 0.01$; female $r = -0.577$, d.f. $= 13$, $p = <0.05$; male $r = -0.559$, d.f. $= 12$, $p = <0.05$), demonstrating that their range increases dramatically from 465 km^2 with 453 mm of rain to 2525 km^2 with 184 mm of rain. It may well be the case that elephants are at the limit of their range in areas receiving <150 mm of rain, since they would be unable to cover sufficient ground to satisfy their food requirements at rainfall levels below this figure.

We have already noted that there is a significant relationship between rainfall and NAAP, so it is not surprising that Coe *et al.* (1976) have demonstrated that large herbivore standing crop biomass (log$_{10}$ kg/km^2) may be reliably predicted from measures of precipitation (log$_{10}$) in areas receiving less than 700 mm of rain per annum. This expression, derived for 20 wildlife areas in east, central and southern Africa is

$$y \text{ (large herbivore biomass)} = 1.685(\pm 0.238 \times -1.095(\pm 0.661)$$

where $x = \log_{10}$rainfall (mm/a)
$\quad (r = 0.96, \text{d.f.} = 19, p = <0.001)$

This relationship holds for all the 20 areas included in the regression, but four other areas were excluded from the computation since two of them (Lake Manyara and Amboseli) were areas of lush vegetation and abundant ground-water, while the other two (Rwindi Plain and Ruwenzori) were located on eutrophic volcanic ash and alluvial soils, which accounted for the higher large herbivore biomasses recorded for these locations (Figure 8.1). It is worth noting that the large herbivore biomass in 12 wildlife areas (included in the regression above) in semi-arid regions receiving <700 mm/a were also highly correlated ($r = 0.94$, d.f. $= 10$, $p = 0.001$) (Coe *et al.*, 1976).

Bell (1982) has analysed further rainfall/biomass data in which he has explored the effect of soil nutrient status on this large herbivore/rainfall relationship and the manner in which it may be modified by water availability,

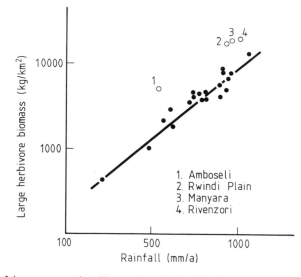

Figure 8.1   Linear regression illustrating the 'optimal carrying capacity' for the large wild herbivore standing crop biomass and mean annual rainfall, for twenty African wildlife areas. The four points above the line represent those conservation areas not included in the regression, which either receive added groundwater or occur on fertile volcanic soils (see text) (data after Coe *et al.*, 1976)

confirming that many areas which show elevated large herbivore biomasses do so on soils with higher nutrient levels. East (1984) recently demonstrated that a similar relationship can be observed for 19 out of 23 individual species in arid/eutrophic savannas, suggesting that natural populations as well as communities of large savanna mammals tend to be close to the limits set by their food resources.

If we return to our observation that the production/biomass ratio of vegetation decreases from deserts to forests, we see that the amount of material immobilized in the plant standing-crop biomass is higher in forests. We also observe that in the more productive savannas the larger mean body size of large mammals in these areas (Coe, 1980b; Western, 1980) will result in a higher P/B ratio for the animals comprising these communities. Coe *et al.* (1976) and Coe (1982) calculated that the P/B ratios for large herbivores should be about 0.05(5%), 0.20(20%) and 0.35(35%) for animals weighing respectively >800 kg, 100–750 kg and 5–90 kg. Comparatively few figures are available in the literature from which the relationship between the P/B ratio and body size can be calculated. Houston (1979), however, has estimated mortality for adults and first year young for fifteen species of Serengeti large herbivores. In the absence of more accurate data on secondary production, if we equate annual biomass mortality with secondary production, on the assumption that the populations are relatively stable from year to year in the absence of some large climatic perturbation, we observe that body size is a good predictor of their 'turnover rate' ($r = -0.886$, d.f. $= 14$, $p < 0.001$) (Coe, 1982). It therefore appears that a large herbivore's mean body size is related to the prevailing climate and the habitat type (Coe, 1980c). In semi-arid environments it may be suggested that these parameters are in turn related to the rate of nutrient circulation, a feature imposed by the vegetation growth form and rate of turnover. Large mammal communities are characterized by multiple equilibria, the temporal elements of the environment altering conditions in a manner that favours the short-term increase or decrease of some components of these communities. Under these variable conditions the numbers of individual species may change with time, although the overall biomass of large herbivores that may be sustained under the prevailing conditions is determined by the rainfall and its effect on primary production.

Although we have seen above that the species component of large mammal communities may be partly separated by body size, spatial and temporal elements of their environment, or even by the behaviour of individual species (Jarman, 1974), it is also clear that strong competitive effects are present even if they are only mediated through a small part of a much larger population. On the Serengeti Plains, zebra, wildebeest and Thomson's gazelle all take part in a clockwise migration around the grasslands, a movement pattern that takes advantage of differing regimes of production due to geographical variations in climate. Bell (1971) and Gwynne and Bell (1968) suggested that there existed a grazing succession between these three herbivore species, in which the zebra ate the larger coarse grasses, the wildebeest ate the longer thin-walled grasses, while the Thomson's gazelle ate the short grass remains on well-grazed sward. Hence a type of grazing facilitation was proposed in which the feeding habits of the larger broad-mouthed species stimulate short growth which provides more suitable forage for the more slender-muzzled Thomson's gazelle. McNaughton (1979) has shown that there is a significant association of Thomson's gazelle with areas previously grazed by wildebeest. This association facilitates energy flow through the gazelle population as a consequence of heavy grazing by wildebeest (Delany and Happold, 1979). Maddock (1979) has cast doubt on the grazing facilitation hypothesis by demonstrating that although zebra and wildebeest are closely associated in the dry season they do not follow the same migration paths. She suggests that the coexistence of these two species is due to their use of feeding refuges, though McNaughton (1979) considers that such an effect is more marked in some areas of the Serengeti vegetation mosaic than others. Thus although Sinclair (1979) concludes, quite reasonably, that it is unlikely that any herbivore is entirely dependent on another for its survival, evidence is accumulating to indicate that wide spectra of wild herbivore species are far more interdependent than has hitherto been supposed.

We have thus observed that natural communities of wild mammals strike a remarkable degree of natural balance through the acquisition of a wide variety of evolutionary traits that allow communities to maintain high levels of diversity and biomass in conditions of variable climate and their concomitant effects on primary production. It is therefore of value to examine the way in which ecosystems function when dominated by pastoralists and their domestic stock.

ARID ZONE CARRYING CAPACITY AND PASTORALISM

Cattle, sheep and goats were first domesticated about 11 000 years ago, followed by the camel and donkey some 5000–7000 years later (Boughey, 1971). Their arrival in Africa dates from about 7000 BC (Epstein, 1971; Cloudsley-Thompson, 1977) when several waves of migration took place, during which the humpless long-horned varieties and the later humpless short-horned forms gave rise to the modern trypanotolerant varieties (Cumming, 1982). Movement across the Sahel and the Sahara took place quite early, but further movement to the south was blocked by a broad belt in which the cattle trypanosomiasis vector, the tsetse fly, prevented further migration. Cumming (1982) considers that the first movement of Sanga cattle by the Hottentots, through the tsetse belt into southern Africa, did not take place until about 1000 years ago from East Africa. Stiles and Muro-Hay (1981) dated material from stone cairns or 'galla graves' in Kenya's Northern Province, which revealed that some of the contents were up to 3500 years old, suggesting that the movement of people (probably with domestic animals) into East Africa is somewhat earlier than has previously been supposed. Indeed, it may well be that evidence which suggests that the climate of the Sudan began to get drier about 5000 years ago, when camels begin to replace cattle (Cloudsley-Thompson, 1977), may well have been an important feature that began to drive people to the south. Whatever the time span that is finally agreed upon for the presence of pastoral people in the Sahel and areas further south in Africa, the drastic environmental changes that have attended their presence are abundantly apparent, and represent one of the most important ecological problems demanding an urgent solution.

Since the very nature of pastoralism is to seek new pasture for domestic stock it is no surprise to observe that most of the peoples of east and central Africa have only occupied their present ranges within the last 400–1000 years (Ogot, 1981). We can therefore hardly be surpised when we learn that many of these peoples have no folk memories of the people that occupied the land before they arrived, referring as the Turkana do in north-western Kenya to a land unoccupied and well vegetated when they first arrived 200–250 years ago, a stark contrast to the picture that prevails today. In many respects this picture of migration, exploitation, overexploitation and further migration is remarkably similar to the scenario proposed by Martin (1973) to account for the loss of 70% of the Pleistocene Megafaunal genera following the arrival of man in the New World 12 000–13 000 years ago. A central problem in any discussion of pastoralism is to decide whether the apparent overstocking of domestic herds above the carrying capacity of the local environment is related to their own concept of wealth, expressed in terms of the numbers of animals the local environment will support, or whether they simply increase their animal numbers in response to an increase in their own numbers.

Brown (1971) considered the ecology of pastoralism (with particular reference to the Masai of Kenya and Tanzania) by asking the question 'can the widespread overgrazing and erosion in pastoral areas . . . be attributed to "prestige" overstocking' or could it be explained in more ecological terms. One of the most important features of these peoples' lives is their reliance on a renewable protein resource—milk—and to a much less common extent for many other pastoral people—blood, provided by their domestic stock. Since the condition of these animals is closely related to the condition of their range and the effect of rainfall on primary production, it is immediately apparent that milk production will be abundant in the short and infrequent wet seasons, while at the height of the dry season, or during extended periods of drought, milk production will be minimal. When milk is the main item of diet, the number of animals that are maintained must take account of the fact that a cow's lactation period only lasts about six months, so in consequence a family unit will need to maintain twice that number of animals to provide their year-round needs. Examining the number of 'standard stock units' required to support an average family unit, in relation to the whole population and the area of land required to support them, Brown (1971) concluded that it was the increase in human numbers and the nature of the pastoral system that has led to problems of overgrazing, land degradation and the attendant problems of a drastically reduced carrying capacity in arid Africa. It did not seem to be related to the more common explanation that pastoral people keep excessively large herds as part of a cultural system based on the notion of the prestige value of a large number of cattle. In many areas today, though, very large herds are commonly being herded by pastoralists on behalf of a wealthy 'third person', when the degree of overstocking is very high, and bears little relation to tribal cultural factors.

During their study of the relationship between the biomass of large herbivores and rainfall, Coe *et al.* (1976) included in their regression data which included wildlife and domestic stock in pastoral areas, collected by Watson (1972) (Figure 8.2). These six points whose annual rainfall lay between 218 and 710 mm a^{-1} all lay above, and almost parallel to, the 'optimal carrying capacity' line for wildlife, suggesting that pastoralists in at least these regions maintained their domestic stock above the carrying capacity. The biomasses used from these areas were calculated by Watson (1972), using slightly higher

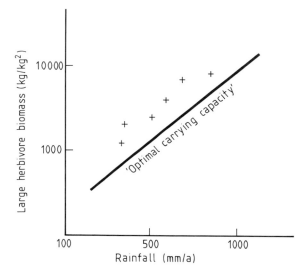

Figure 8.2   Data for the standing crop biomass of large herbivores (+) on six pastoral areas (mainly domestic stock) and mean annual rainfall, in relation to the 'optimal carrying capacity' derived for wild herbivores occupying twenty conservation areas. Note that a regression for these pastoral ecosystems would lie parallel to and consistently above that for wildlife (see Figure 8.1) (data after Coe *et al.*, 1976)

figures than those employed by Coe *et al.* (1976), but since several of these areas suffered drastic mortality in the severe drought of 1973–4 (Casebeer and Mbai, 1974) it seems highly likely that they were overstocked. Data for ten pastoral areas in Kenya are given by Dyson-Hudson (1980) in respect of domestic stock numbers only, but using the unit weights of Coe *et al.* (1976) we observe that a significant linear correlation exists between total domestic stock biomass and rainfall ($r = 0.840$, d.f. $= 9$, $p = 0.01$). Rainfall was estimated from Kenya meteorological data (Anon, 1976). If we examine the correlation between individual components of the domestic stock biomass we observe that a similar significant correlation exists between cattle biomass and rainfall ($r = 0.780$, d.f. $= 9$, $p = <0.01$). Camels, however, were not significantly correlated with rainfall ($r = -0.007$, d.f. $= 9$, NS) although this feature was strongly influenced by the Isiolo data which was unusually high, a factor that may be influenced by the fact that this area is an important trading area for camels. Sheep and goats were poorly correlated for similar reasons ($r = 0.366$, d.f. $= 9$, $p = >0.1$). Phillipson (1977), in an analysis of national cattle figures for Kenya, concluded that deviations from a simple linear pattern, especially at rainfall levels <208 mm a^{-1}, were due to climatic vagaries of these environments, the incidence of tsetse flies and local cultural factors. A similar exercise was carried out by Bourn (1978) in Ethiopia and on a continental scale for cattle numbers, when he compared his data with the suggested 'optimal carrying capacity' (OCC) levels of Coe *et al.* (1976). He showed that in Ethiopia at the lower rainfall levels the biomasses of cattle exceeded the line for tsetse-free areas, while from about 1000 mm of rain his regression lies below the OCC line. Cattle biomass data for the whole continent indicate that the regression for tsetse-free areas lies consistently below the OCC line. It is important here to note that these are cattle figures, which do not take account of the elevated biomasses expected at low rainfall due to the addition of camels and goats, while at higher rainfall a larger percentage of the land is being used for agriculture. Watson (1969), in a detailed aerial survey of South Turkana, Kenya, indicated that the ratio of cattle, camels, sheep and goats varied according to the local physiography and climate, which suggests that the stock carried by pastoralists can be governed by what the local climatic conditions will allow rather than solely to ingrained cultural mores.

We have observed above that arid ecosystems manipulated by pastoralists seem to be consistently maintained above the optimal carrying capacity, whether the explanation of this phenomenon is social, political or ecological (Lamprey, 1983). Sahelian famine has exercised the minds of meteorologists, geographers and ecologists for many years, who have sought to explain drought conditions in terms of the long-term variation in rainfall patterns (see Chapter 1). After a period of drought in the early 1970s, there followed a short interval of higher precipitation, though even here they were barely above the average values for the period 1931–60 (Grove, 1986). Bryson (1975), Lamb (1982) and Winstanley (1973), however, considered that the low rainfall of the early 1970s was part of a general trend to increasing aridity. Tending to support this view is the observation that, in the early 1980s, an area from Senegal to the Tigrai region of Ethiopia and the Red Sea Hills experienced a reduction in rainfall 60% below the 1931–69 mean (Grove, 1986). Yet in spite of these trends much of the western Sahel has experienced rainfall levels in 1988 which exceed anything seen in the last 20 years (Toumalin, 1988). Such observations can be supported at many levels, but especially at that of the consistently reduced carrying capacity of the Sahelian zone. Their explanation, however, requires something more than long-term drought cycles, which are outside our human control. There is now ample evidence to suggest that climatic cycles of high and low rainfall do exist in East Africa (Laws, 1965a; Phillipson, 1975) and the Indian Ocean (Stoddart and Walsh, 1979), in which we may observe 10 year, 40–50 year and even longer-term cycles. Whether these phenomena are truly cyclic and hence of predictable periodicity, or simply random events (Burroughs, 1986), is quite another matter. We may conclude that the problem of the drought-ravaged lands of Africa may at least in part be explained in ecological terms, of natural and anthropogenic effects, as well as the purely physical effects of climatic perturbations.

The factors responsible for the deterioration of the rangelands are complex and in many cases do not necessarily relate directly to the activities of pastoralists and their domestic stock. As human populations of agricultural peoples have increased at rates often in excess of 4%, so they have extended their range from areas of high rainfall to climatically marginal land at lower altitude, where in spite of frequent crop failure, they are still utilized and as a result bring them into territorial conflict with the pastoralist. In some cases activities that might be expected to assist the pastoralist may in others have the opposite effect, for Ormerod (1976) has suggested that in regions of the western Sahel tsetse-eradication programmes have assisted the southward extension of the arid lands, when cattle and other domestic animals have been introduced into these sensitive environments following fly removal.

The research of the wildlife biologist may often seem somewhat remote from the needs of human beings attempting to scratch a living from the arid savannas, yet recent work by Sinclair and Fryxell (1985) has indicated that the behaviour of migratory large herbivores provides an excellent analogue for understanding the problems of the Sahel. Over a period of 20 years Sinclair and his colleagues at the Serengeti Research Institute have been studying the ecology of these mammals on the Serengeti Plains, where they have discovered that they have evolved two major behavioural patterns to cope with the rigours of a semi-arid environment, especially in respect of the seasonal limitation of food and water supplies. Some truly arid-adapted species are physiologically independent of standing water sources, while others carry out small-scale migratory movements to exploit patchy food resources. In respect of the other, usually larger grazing, herd-dwelling species in the southern Sudan, East Africa and Namibia, their response to intensely seasonal rainfall and grazing is to migrate large distances to take advantage of varying food availability. On the Serengeti Plains up to 2 million white-bearded wildebeest have been studied in great detail (Maddock, 1979; McNaughton, 1979; Sinclair, 1983). These studies have revealed that the wildebeest move annually from tall grasslands of low protein value in the dry season, where rainfall of up to 1000 mm a^{-1} may be expected, to areas of short grass in the wet season on the open plains, although the actual rainfall of these habitats may be marginal ($c.400$ mm a^{-1}). In spite of the apparent aridity of the short-grass plains they support a seasonally high ungulate biomass, when the animals can replenish their fat reserves and give birth to their young (Sinclair, 1977; Watson, 1969), before they once more depart for the long-grass regions when the short-lived rain pools dry up. This movement strategy therefore allows the plains ungulates of Serengeti (and the white-eared kob of the southern Sudan) to maintain much higher population densities than they could sustain if they remained in one area throughout the year.

When we examine the behaviour of pastoral peoples and their domestic animals, there is a striking similarity between them and the large migratory wild herbivores (Franke and Chasin, 1980; Lamprey, 1983; Breman and de Wit, 1983; Sinclair and Fryxell, 1985). In many respects one might almost envisage that the initial evolution of pastoral migratory patterns has developed through the nomads following, rather than directing, the movements of their animals (Krebs and Coe, 1985). In much of the western areas of the Sahel, pastoralists undertake movement patterns in which they spend short periods in the north to take advantage of the high-quality grasses available there, before they turn southwards to higher rainfall grasslands. These movements are conditioned by the availability of standing water and forage (Bremen and de Wit, 1983). In the south, however, the pastoralists come into conflict with millet, sorghum and groundnut cultivators, who under increasing population pressures have extended their agricultural ranges, although the 'symbiotic' provision of dung and urine does help to enhance the otherwise poor soil fertility (Sinclair and Fryxell, 1985). These general patterns of behaviour are very ancient, and have probably changed very little since they were first established in the Sahel at least 5000 years ago (Dumont, 1978). Thus it is evident that the severe environmental damage that had been observed in the last 50 years is a comparatively recent event, the cause of which seems much more likely to be found in the complexities of increased anthropogenic intervention, and medical and veterinary care of the people and their animals, rather than in long-term climatic changes and associated natural phenomena.

Sinclair and Fryxell (1985) have convincingly argued that the considerable increase in Western aid over the last 25 years may well have had a strong influence on these drastic patterns of advancing desertification. They underline the active governmental encouragement of cultivators to move their operations deeper into the grazing range of nomadic pastoralists. As a consequence these people have begun to lose much of their traditional and ecologically essential southern grazing lands, while the assumption that the major problem faced by pastoralists was a lack of water for their stock led to the establishment of additional boreholes and water supplies (Wade, 1974; Franke and Chasin, 1980, 1981). These events closely paralleled the attainment of independence in the early 1960s by many of these states, after which the former ranges of many of the pastoral peoples were disrupted and restricted by the newly created political boundaries, leaving the pastoral peoples little alternative but to settle in much more restricted ranges around their newly constructed water supplies (Wade, 1974). These new political structures and the increased availability of veterinary and medical services then resulted in dramatic increases in the numbers of animals and people in their now greatly restricted range, leading inevitably to severe local over-grazing and soil degradation around the newly established water points, which spread and then coalesced, until the carrying capacity of much of their grazing land was dramatically reduced (Sinclair and Fryxell, 1985).

Here we have observed just a few of these important events which clearly indicate that the Sahel is in the grip of a man-made famine (Krebs and Coe, 1985), which does not of course mean that there are no regular or random cycles in rainfall in this region, only that these physical effects are of minor importance compared with the near-synergistic consequences of modern anthropogenic agencies. Sinclair and Fryxel (1985) point out that evidence for their settlement-overgrazing hypothesis is supported by the fact that ranches on which minor local migratory patterns are still encouraged do not suffer the same levels of damage as those in surrounding areas with reduced movement.

The support of these views would be even stronger if it could be shown that the man-induced events in the Sahel have actually led to a reduction in rainfall, so that as the overgrazing gets worse so local precipitation is reduced still further. Contentious though such arguments may seem, Sinclair and Fryxell (1985) argue with convincing and persuasive force that such evidence is available. They suggest that in areas of increased settlement, the reduced vegetation cover and higher soil surface albedo (the degree of reflectance of radiation from a surface) result in cooler soil and decreased rainfall, while in areas where migratory patterns are still encouraged, the intact plant cover results in a decreased surface albedo and warmer soil, greater generation of thermals and increased rainfall (Figure 8.3).

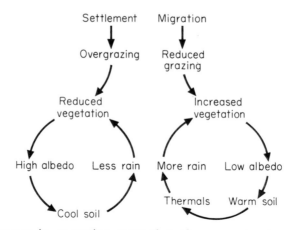

Figure 8.3 Changes in vegetation cover through overgrazing in settled pastoral ecosystems may lead to reduced rainfall through a feedback system involving a high albedo, while well-vegetated sites where migration is still encouraged generate greater rainfall through the lower albedo and warmer soils (after Sinclair and Fryxell, 1985)

The above evidence, which has been generated from air circulation models (Charney, 1976), although by no means yet completely proven, seems to fit the notion of an important climatic feedback process, which accords with the ecological observations described above. Sinclair and Fryxell (1985) do, however, point to one very important consequential warning note, which is that short-term famine aid programmes stand little chance of success unless they are carried out in conjunction with soundly based ecological principles, some of the directions of which have been indicated in a seemingly unrelated study of wildlife, a field hardly likely to generate extensive funds from aid agencies concerned primarily with human welfare problems in the Sahel. The ecological analogues of the migratory wildlife and pastoral systems are very strong, which we will ignore at our peril if we are ever to learn or devise the most effective means of utilizing the world's arid lands.

ARID LANDS FOR WILDLIFE AND/OR DOMESTIC STOCK?

We have observed above that the conservation areas of semi-arid Africa seem capable of maintaining their populations of large mammals at something close to their carrying capacity, while pastoral peoples appear to adopt a strategy of overexploitation. In recent years, however, increasing human populations have not only placed direct pressure on pastoral environments but have also imposed indirect compression effects on National Parks and Reserves. These effects have been the subject of often acrimonious debate between the interventionists and the non-interventionists, who have respectively held the views that the larger species should be culled or be left to reach their own form of natural balance (Caughley, 1976; Corfield, 1973; Laws, 1969b; Phillipson, 1975; Malpas, 1978). At the time that these debates took place it was held that the carrying capacity of elephants was about $0.5\,\mathrm{km}^{-2}$ in savanna environments, while in the Tsavo National Park, Kenya, these mammals had reached nearly $1.5\,\mathrm{km}^{-2}$ as development and land pressures drove them to seek shelter in the conservation area (Laws, 1969a), as a result of which woodlands were destroyed and converted to grasslands in Tsavo and over much of savanna Africa. Since this time, such arguments seem somewhat erroneous, for it is difficult to believe that the populations of African elephants (and black rhino) have been reduced by over 80% through poaching since the early 1970s (Parker and Amin, 1983; Douglas-Hamilton, 1987). Indeed, it is no exaggeration to suggest that the African elephant and black rhino will be extinct over most of their range in the next ten years unless immediate remedial measures are taken to protect them. To the people hungry for a small plot of land on which to scratch a living either with domestic stock or by subsistence agriculture, the imminent disappearance of large mammal species is probably of little importance, but to the national exchequer, their foreign exchange value as a tourist resource are of immense importance (Eltringham, 1984; Myers, 1972).

Such arguments, based solely on the conservation ethic, however, frequently fail to take into account that wild animals have traditionally comprised, and still do, a large part of the protein diet of many humans in virtually all tropical environments. We now know a great deal about the organisms that are utilized as food in forest environments of West Africa and South America, where the wide dietary spectrum comprises virtually everything from arthropods (insects and crustaceans) to mammals. Asibey (1974, 1977) reports that in some parts of Ghana up to 73% of the locally produced meat comes from small mammals such as the cane rat, hare and giant rat, while additional items regularly eaten range from small reptiles and amphibians to insect larvae and the giant African land snail. The importance of these local sources of protein is well illustrated by a single small 'chop bar' (restaurant) at Sunyani, Ghana, where up to 360 people were fed on bushmeat in a single day (Asibey, 1977). This high level of utilization in forest environments is summarized by Ajayi (1979) who estimated, from a number of sources, that wild animal protein comprised 70–80% in Cameroon, 20% in Nigeria, 73% in Ghana, 70% in the Ivory Coast and 80–90% in Liberia, while Curry-Lindahl (1969, 1972) estimates that up to 50% of the local diet in the Congo (Brazzaville) is composed of locally caught bushmeat. These figures are similar to those available for South America, where Dourojeanni (1985) has calculated that 87–90% of the domestic meat consumed in the Peruvian Amazon is derived from animals caught in the wild (Kyle, 1987).

Clearly the items taken locally will depend on the habitats available for hunting and their conservation status, for Charter (1971) records that in southern Nigeria fish made up 60% of the local protein intake, domestic animals 21% and wild mammals 19%. The economic value of this resource is often underestimated but when we observe that the value of this wild resource in Southern Nigeria between 1965 and 1966 was up to $20 million, it is clear that this is a significant local economic factor. In Ghana it may be estimated that during an average month between 1968 and 1970 up to 20 469 kg of bush meat valued at $9411 was offered for sale in Accra (Asibey, 1974).

It is interesting to note that these high levels of utilization are for forest environments, where the mean size of herbivores that are potentially available as food are much smaller than those of the African savanna. This apparent anomaly is explained by the fact that most of the nutrients of a forest are immobilized in wood, while in the more open habitats of the semi-arid and arid grasslands and woodlands the vegetation turns over more rapidly and as a consequence more nutrients are tied up in the bodies of large mammals (Coe, 1980a). Over much of savanna Africa the sale of game meat is either very strictly controlled if not banned altogether, so that although these sources of protein must have been of paramount importance in the past, today it is much less significant. Richter (1969) in a study of the people of the arid Kalahari region of Botswana has calculated that they derived up to 16.5 kg per individual per acre from game animals between 1965 and 1967. Individual groups of hunter-gatherers in the Kalahari, whose ecologically sound life-styles are now threatened by development, have been calculated to derive 273 gm/d of meat or 68 gm/d of protein from their hunting activities (Eltringham, 1984; Silberbauer, 1981). These hunting lifestyles are now much less common than they were in the comparatively recent past, although in many areas the smaller species are still abundantly available for food and the artificialy restricted ranges provided in many National Parks and Reserves show extensive signs of overpopulation. Indeed, it is almost certainly true that in some countries, where hunting on controlled areas has been banned for some years as a conservation measure, the level of poaching (illegal hunting) has increased dramatically in the absence of well-organized (and armed) sport hunting parties.

The value of tourism to African countries with large National Parks and spectacular habitat diversity to provide visual variety, both in terms of wildlife and scenery, is clearly of great economic importance, but the cost of main-taining such facilities are high and their susceptibility to even short-term political instability makes it dangerous to rely on such income as a major source of foreign exchange. Eltringham (1984) has discussed the economic and social problems that arise from placing undue, long-term reliance on tourism, and suggests like other wildlife biologists and economists (Pullan, 1984) that the survival of these important conservation areas for their aesthetic, scientific, educational and genetic value may only be assured if they receive much greater financial support from the international com-munity.

We are therefore faced with the very difficult problem of devising a development strategy for savanna Africa and similar environments else-where, which will provide additional, rather than alternative, forms of land use for arid habitats. At moderate altitude land may be effectively used for agriculture and the ranching of domestic livestock, but on more marginal land below these levels it seems highly probable that wildlife and/or domestic stock may provide a more ecologically sound form of land use. Indeed, as the patterns of land ownership change from communal to individual title deeds the future of wildlife outside conservation areas will be in the hands of these new individual land owners (Coe, 1980b; Parker and Graham, 1971). Thus the future of both wild species and the pastoral lands must be viewed together, as an integrated form of land use to prevent further ecological degradation, while at the same time demonstrating that a new or modified system of land use is economically advantageous.

Wildlife may be utilized in three very different ways:

(a) management and utilization of free-ranging wild species on a sustained yield basis, as a source of meat, for sport hunting, for the capture and sale

of live animals for zoos or other wildlife areas (either official or private), or for their other products such as skins, hooves and horns;
(b) the selection of a range of wild herbivorous species for enclosed 'game farm' operations;
(c) the domestication or perhaps more correctly semi-domestication, since the former implies 'bringing an animal into subjection to or dependence on man' (Skinner, 1973), a difficult and time-consuming process.

The last of these forms of wildlife utilization need not concern us here for they do not seem to provide an alternative style of land use for the peoples of the arid lands. Experiments, however, on the Galana Ranch in Kenya with buffalo, eland and oryx have demonstrated that the last-named species, which is so well adapted to dry habitats, can be handled, managed and marketed (Field, 1975; Lewis, 1977, 1978; Stanley-Price, 1976), while in Zimbabwe and South Africa the eland has been the subject of similar experiments (Bigalke and Neitz, 1954; Huygelen, 1955; Jelliman, 1913; Kerr and Roth, 1970; Kerr *et al.*, 1970).

Game farming operations using a variety of species on enclosed land provide an alternative form of land use either alone or in conjunction with domestic stock. Hopcraft (1975) has demonstrated that a ranch in eastern Kenya can be effectively converted from a cattle to a game ranch. He concludes that from an economic standpoint the cost effectiveness of wildlife is nearly eight times greater in terms of net income than cattle, but specially because the cost of maintaining cattle takes 70% of the gross income against only 20% for wildlife (Hopcraft, no date). It is here difficult to understand why local governments do not take the ecological and economic significance of this type of research much more seriously. Such ranching techniques are now widely employed in Botswana, South Africa and Zimbabwe, especially on land whose value had already been severely depressed through damage by domestic stock. The main species used in these operations are blesbok, eland, impala, springbok and greater kudu, where even this small spectrum provides a good balance of grazers, mixed feeders and browsers. It is perhaps interesting to note that were it not for their maintenance on ranches, the blesbok, bontebok and black wildebeest would probably now be extinct. The cost effectiveness of utilizing these animals has been described by a number of authors (Child, 1970; Joubert, 1968; Skinner, 1966, 1967, 1973; von la Chevallierie, 1970), who conclude that although a great deal more work needs to be done on the logistics of game ranching, once established it is far less costly than cattle ranching in terms of disease prevention, fencing and labour.

During the long debate between wildlife biologists and ranchers, the former have often argued that the high dressing percentages of game animals makes them a far more efficient form of range use than that of domestic stock

(Ajayi, 1979; Crawford, 1974; Kyle, 1972; Skinner, 1973; Talbot *et al.*, 1965). It is, however, worth examining the data on which these authors base these assumptions, for when we do so we observe that like so many other life history parameters, dressing percentage (the utilizable proportion of the carcass) is significantly scaled to body size (Table 8.1). Thus we observe that the relationship between the dressing percentage and unit weight of eleven wild herbivore species is barely significant at the level $p = <0.1$ (d.f. $= 10$, $r = -0.545$), while if we add the unit weight of cattle to the wild herbivores this significance is increased to $p = <0.05$ (d.f. $= 11$, $r = -0.575$). A least squares regression of $\log_{10}$ dressing percentage on $\log_{10}$ unit weight showed a higher level of significance (wild herbivores: $r = -0.653$, d.f. $= 10$, $p = <0.05$; wild herbivores + cattle: $r = -0.679$, d.f. $= 11$, $p = <0.02$). Such an analysis does not suggest that this makes wildlife a less valuable source of range use, but rather that the smaller species are often more valuable in relation to the rate of turnover, which as we have already seen is itself scaled to body size. A further argument in favour of meat from wildlife is their fat levels, which lie consistently below 4% compared with 13.7% in an improved Boran bull (Ledger, 1968), based on the assumption that fat is an unnecessary and unhealthy carcass constituent (Crawford, 1974). Such a contention is mainly based on the problems faced by the better-fed members of human society, but may well not be so true for those surviving on a subsistence diet and economy

Table 8.1   Unit weights and dressing percentages of large wild herbivores and cattle

Herbivore	Unit weight (kg)	Dressing percentage (kg)	$\log_{10}$ unit weight (kg)	$\log_{10}$ dressing percentage (kg)
Thomson's gazelle	15	58.00	1.17609	1.76343
Springbok	26	57.90	1.41497	1.76268
Grant's gazelle	40	63.20	1.60206	1.80072
Impala	40	58.80	1.60206	1.76938
Reedbuck	40	55.70	1.60206	1.74586
Blesbok	53	52.90	1.72428	1.72346
Topi	100	53.60	2.00000	1.72916
Wildebeest	123	50.40	2.08991	1.70243
Kongoni	125	52.00	2.09691	1.71600
Great kudu	136	56.60	2.13354	1.75282
Eland	340	52.15	2.53148	1.71725
Cattle	180	52.40	2.25527	1.71933

Unit weights after Coe *et al.* (1976).
Dressing percentages after Skinner (1967) and Talbot *et al.* (1965).
Cattle dressing percentage after Pratt and Gwynne (1977).

(Coe, 1980b), although a lower fat content does obviously produce a carcass with a higher percentage of saleable lean meat.

Arguments in favour or against the utilization of wildlife as a natural resource can only be answered convincingly in economic terms. Recently Child and Child (1986) have analysed data for game ranching areas in Zimbabwe where it has been possible to compare the performance of cattle with game. The Buffalo Range Estate, in the south-eastern lowveld of Zimbabwe, receives an annual rainfall of <500 mm and is the longest established game ranch in the country, on which cattle are grazed on 20 000 ha and game on a separate section of 10 000 ha. The estate's excellent records reveal that the net revenue for the period 1978–84 was Z$0.72 ha^{-1} for game compared with Z$0.10 ha^{-1} for cattle, though when these figures are corrected for inflation over the same period the real prices for game are Z$0.63 ha^{-1} compared with Z$0.18 ha^{-1} for cattle, notably 3.5 times higher for the game.

These figures, however, do not highlight the major differences in revenue which lie in a capital investment of Z$965 000 for the cattle operation compared with only Z$36 000 on the wildlife section. Further data are provided for an area of the Zimbabwe midlands, which receives slightly less marginal rainfall (*c.*650 mm a^{-1}) and the net revenues from a 10 000 ha ranch is estimated as Z$3.78 for cattle and Z$6.35 for game, or expressed as the net revenue per kilogram biomass the respective net revenues were higher, at Z$0.08 for cattle and Z$0.18 for game. The special significance of these data is that they are illustrative of what can be achieved in marginal environments, especially when we learn that as a result of these ranching practices the Buffalo Range Estate has changed from being the most ecologically degraded estate in the early 1960s to the most valuable in 1986 (Child and Child, 1986). In spite of the apparent advantages of utilizing wildlife on marginal land, this form of biological resource use is still far from fully developed, even in Zimbabwe, for as the above authors point out, the skins and their attendant secondary industries are poorly developed, live game sales are under-exploited and the potential for plains game hunting on such estates is only in its early stages of development. The biggest problem though lies on the ecological and the logistic fronts, for there is a lack of information available to game farmers on management techniques necessary for maximising sustainable returns from game in savanna ecosystems and the development of efficient cropping and handling techniques (Child and Child, 1986).

As movement from National Parks and Reserves has become more restricted through human settlement, so the managers of conservation areas have had to face the need for cropping their wildlife populations to prevent local habitat deterioration. These management techniques are best developed in South Africa where animals are cropped under strict controls in National Parks and surrounding areas, as well as on private land. The returns from such operations in South Africa are summarized by Hanks *et al.* (1981), who record that the mean number of animals removed annually from the Kruger National Park and the Natal National Parks between 1975 and 1979 were respectively 2261 and 4637, while cropping permits issued in the Transvaal averaged 6546 a^{-1} and animals taken on private land in the Transvaal in 1977 totalled 7984. Additional wildlife utilized in this period 1975–9 included 3011 animals per annum captured for live sale from the Natal National Parks. In financial terms these operations yielded 263 800R a^{-1} for the Kruger National Park and 221 616R a^{-1} for the Natal National Parks, returns which must be viewed as additions to the parks' tourist income.

CONCLUSIONS

The manner in which wildlife and pastoralism compete is well exemplified by recent studies in the Kora National Reserve (KNR) in eastern Kenya, in a joint Royal Geographical Society and National Museums of Kenya investigation. The intrusion of Somali pastoralists in this area is part of a southward expansion movement that began at least as early as the middle of the nineteenth century, while to the west subsistence agriculturalists have extended their range from the more productive highlands, under the severe land pressure that now exists there. Aerial counts of wild and domestic herbivores of the KNR, conducted by the Kenya Rangeland Ecological Monitoring Unit (KREMU) have revealed that while the biomass of wild herbivores (when the smaller species are included) lie close to the predicted carrying capacity, the presence of between 63.69 and 85.13 domestic herbivores km^{-2} in 1983–4, representing a biomass of 4368.4–8554.1 kg km^{-2}, exceeds the carrying capacity by up to 4.2 times (Olang, 1986; Coe and Collins, 1986). Although it is evident that the pastoralists are short of grazing in all these areas, especially during times of drought, it is equally obvious that their intrusion into conservation areas, which represent such a small percentage of the total available land, will only reduce them to the same degraded state as their surroundings, destroying habitats that are of inestimable value as baseline areas for all future studies of arid zone management and utilization. It is, however, equally clear that management strategies for these protected habitats will only succeed if the local people are involved in their planning and execution. Such needs are equally true of all forms of arid zone development, whether they are related to competition between domestic stock and wildlife, or the development of water holes or irrigation, for in all cases the absence of sound ecological survey and planning will reduce the carrying capacity of these lands still further.

To many of us who have been involved in the study of African wildlife throughout our working lives, it is an unpalatable truth that only by utilizing wildlife, whether it be through cropping for their products or for sport hunting, that the wilderness areas of the continent will survive, by being seen to be able to pay their own way in a regrettably commercial world. To suggest that most of these techniques and operations are directly or immediately applicable to arid regions like the Sahel would be naive, but the ecological message that these arguments transmit are that unless we tackle the human and domestic animal numbers problem with urgency, the image of starving children in the world's arid lands will remain with us as a reminder of our failure.

REFERENCES

Ajayi, S. S. (1979). Utilisation of forest wildlife in West Africa, FO:MISC/79/26, FAO/UN, Rome, pp. 1–76.
Anderson, J. M. and Coe, M. J. (1974). Decomposition of elephant dung in an arid tropical environment, *Oecologia (Berl.)*, **14**, 111–25.
Anon. (1978). *Summary of Rainfall in Kenya for the Year 1976*, East African Meteorological Department, Nairobi, Kenya.
Asibey, E. O. (1974). Wildlife as a source of protein in Africa South of the Sahara, *Biol. Conserv.*, **6**, 32–9.
Asibey, E. O. (1977). Expected effects of land-use patterns on future supplies of bushmeat in Africa South of the Sahara, *Environ. Conserv.*, **4**, 43–9.
Bell, R. H. V. (1971). A grazing ecosystem in the Serengeti, *Sci. Am.*, **224**(1), 86–93.
Bell, R. H. V. (1982). The effect of soil nutrient availability on community structure in African ecosystems, in *Ecology of Tropical Savannas* (Eds. B. J. Huntley and B. H. Walker), pp. 193–216, Springer-Verlag, Berlin.
Bigalke, R. C. and Neitz, W. O. (1954). Indigenous ungulates as a possible source of new domesticated animals, *J. S. Afr. Vet. Assoc.*, **25**, 45.
Blueweiss, L., Fox, V., Kudzma, V., Nakeashima, D., Peters, R. and Sams, S. (1978). Relationships between body size and some life history parameters, *Oecologia (Berl.)*, **37**, 357–72.
Boughey, A. S. (1971). *Man and Environment*, Macmillan, New York.
Bourn, D. (1978). Cattle, rainfall and tsetse in Africa, *J. Arid Environ.*, **1**, 49–61.
Bremen, H. and de Wit, C. T. (1983). Rangeland productivity and exploitation of the Sahel, *Science*, **221**, 1341–7.
Brown, L. H. (1971). The biology of pastoral man as a factor in conservation, *Biol. Conserv.*, **33**, 93–100.
Bryson, R. A. (1975). The lessons of climatic history, *Environ. Conserv.*, **2**, 163–79.
Burroughs, W. (1986). Randomness rules the weather, *New Scient.*, **111**(1516), 36–40.
Casebeer, R. L. and Mbai, H. T. M. (1974). Animal mortality 1973–74: Kajiado District, UNDP/FAO Wildlife Management Project, Project Working Document 5, Nairobi, Kenya.
Cassidy, J. T. (1973). The effect of rainfall, soil moisture and harvesting intensity on grass production on two rangeland sites in Kenya, *E. Afr. Agric. For. J.*, **39**, 26–36.
Caughley, G. (1976). The elephant problem—an alternative hypothesis, *E. Afr. Wildl. J.*, **14**, 265–83.
Charney, J. P. H. (1976). Reply to Ripley (1976), *Science*, **191**, 100–2.
Charter, J. R. (1971). Nigeria's wildlife: a forgotten asset, in *Wildlife Conservation in West Africa* (Ed. D. C. Happold), 60 pp, IUCN Pub. N.S. 22.

Child, G. (1970). Wildlife utilisation and management in Botswana, *Biol. Conserv.*, **3**, 18–22.

Child, B. and Child, G. (1986). Wildlife, economic systems and sustainable human welfare in semi-arid rangelands in Southern Africa, in *Watershed Management in Arid and Semi-Arid Zones of the Southern African Development Coordination Conference (SADCC)*, FAO/Finland Workshop, Maseru, Lesotho (mimeographed).

Cloudsley-Thompson, J. L. (1977). *Man and the Biology of Arid Zones*, Arnold, London.

Cobb, S. M. (1976). The distribution and abundance of the large herbivore community of Tsavo National Park, Kenya, D.Phil. Thesis, University of Oxford.

Coe, M. (1972). Defaecation in African Elephants (*Loxodonta africana africana* Blumenbach), *E. Afr. Wildl. J.*, **10**, 165–74.

Coe, M. (1977). The role of termites in the removal of elephant dung in the Tsavo (East) National Park, Kenya, *E. Afr. Wildl. J.*, **15**, 49–55.

Coe, M. (1980a). African mammals and savanna habitats, *International Symposium on Habitats and Their Influences on Wildlife*, pp. 83–109, Endangered Wildlife Trust, Pretoria, South Africa (mimeographed).

Coe, M. (1980b). African wildlife resources, in *Conservation Biology* (Eds. M. E. Soule and B. A. Wilcox), pp. 273–302, Sinauer Assoc., Sunderland, Mass.

Coe, M. (1980c). The role of modern ecological studies in the reconstruction of palaeoenvironments in sub-Saharan Africa, in *Fossils in the Making* (Eds. A. K. Behrensmeyer and A. P. Hill), University of Chicago Press.

Coe, M. (1982). Body size and the extinction of the Pleistocene Megafauna, *Palaeoecol. of Afr.*, **13**, 139–45.

Coe, M. (1983). Large herbivores and food quality, in *Nitrogen as an Ecological Factor* (Eds. A. McNeill and I. H. Rorison), pp. 345–68, Blackwells Scientific Publishers, Oxford.

Coe, M. (1986). Ecology and development planning, *M.O.A. International Symposium, North–South Dialogue*, Oxford (mimeographed).

Coe, M. (1987). Unforseen effects of control, *Nature*, **327**(6121), 367.

Coe, M. and Carr, R. D. (1983). The relationship between large ungulate body weight and faecal pellet weight, *Afr. J. Ecol.*, **21**, 165–74.

Coe, M. and Collins, N. M. (Eds.) (1986). *Kora: An Ecological Inventory of the Kora National Reserve, Kenya*, Royal Geographical Society, London.

Coe, M. and Kingston, T. J. (1988). The ecology of dung beetles in an African xeric environment, in *Ecophysiology of Desert Mammals* (Eds. P. Ghosh and I. Prakash), Scientific Publishers, Jodhpur, India (in press).

Coe, M. J., Cumming, D. H. and Phillipson, J. (1976). Biomass and production of large herbivores in relation to rainfall and primary production, *Oecologia (Berl.)*, **22**, 41–54.

Colinvaux, P. (1980). *Why Are Big Fierce Animals Rare*, George Allen and Unwin, London.

Corfield, T. F. (1973). Elephant mortality in Tsavo National Park, Kenya, *E. Afr. Wildl. J.*, **11**, 339–68.

Crawford, M. (1974). The case for new domestic animals, *Oryx*, **12**, 351–60.

Cumming, D. H. M. (1982). The influence of large herbivores on savanna structure in Africa, in *Ecology of Tropical Savannas* (Eds. B. J. Huntley and B. H. Walker), pp. 217–45, Springer-Verlag, Berlin.

Curry-Lindahl, K. (1969). Report to the Government of Liberia on conservation, management and utilisation of wildlife resources, IUCN Pub. Ser. Supp. Paper 24.

Curry-Lindahl, K. (1972). *Conservation for Survival: An Ecological Strategy*, William Morrow, New York.

Delany, M. J. and Happold, D. C. D. (1979). *Ecology of African Mammals*, Longman, London.

Douglas-Hamilton, I. (1987). African elephant population trends and their causes, *Oryx*, **21**(1), 11–24.

Dourojeanni, M. J. (1985). Over-exploited and under-used animals in the Amazon region, in *Amazonia* (Eds. G. T. Prance and T. E. Lovejoy), pp. 419–33, IUCN and Pergamon, Oxford.

Dudley-Stamp, L. and Morgan, W. T. W. (1973). *Africa: A Study in Tropical Development*, 3rd ed., John Wiley, New York.

Dumont, H. J. (1978). Neolithic hyperarid period preceded the present climate in the Sahel, *Nature*, **274**, 356–8.

Dyson-Hudson, N. (1980). Strategies of resource exploitation among East African pastoralists, in *Human Ecology of Savanna Environments* (Ed. D. R. Harris), Academic Press, London.

East, R. (1984). Rainfall, soil nutrient status and biomass of large African savanna mammals, *Afr. J. Ecol.*, **22**, 245–70.

Eltringham, S. K. (1984). *Wildlife Resources and Economic Development*, John Wiley, Chichester, UK.

Epstein, H. (1971). *The Origin of Domestic Animals in Africa*, Africana Publ. Corp., New York.

Field, C. R. (1972). The food habits of wild ungulates in Uganda by analyses of stomach contents, *E. Afr. Wildl. J.*, **10**, 17–42.

Field, C. R. (1975). Climate and food habits of ungulates on Galana Ranch, *E. Afr. Wildl. J.*, **13**, 303–20.

Foster, J. B. and Coe, M. J. (1968). The biomass of some animals in the Nairobi National Park, *J. Zool. Lond.*, **155**, 413–25.

Franke, R. W. and Chasin, B. H. (1981a). *The Seeds of Famine*, Allanhead, Osmun and Co., Montclear, N. J.

Franke, R. W. and Chasin, B. H. (1981b). Peasants, peanuts, profits and pastaralists, *Ecologist*, **511**, 156–68.

Glover, P. E. (1970). *Tsavo Research Project Report: June 1968–June 1970*, Kenya National Parks, Nairobi, Kenya (mimeographed).

Glover, P. E. (1973). *Tsavo Research Project: Progress Report for the Year 1973*, Kenya National Parks, Nairobi, Kenya.

Griffiths, J. F. (1976). *Climate and the Environment: The Atmospheric Impact on Man*, Paul Elek, London.

Grove, A. T. (1986). The state of Africa in the 1980s, *Geog. J.*, **152**(2), 193–203.

Gwynne, M. D. and Bell, R. H. V. (1968). Selection of vegetation components by grazing ungulates in the Serengeti National Park, *Nature (Lond.)*, **220**, 390–3.

Hanks, J., Denshman, W. D., Smuts, G. L., Jooste, J. F., Joubert, S. C. J., le Roux, P. and Milstein, P. le S. (1981). Management of locally abundant mammals—the South African experience, in *Problems in Management of Locally Abundant Wild Animals*, pp. 21–55, Academic Press, London.

Harris, D. R. (1980). Tropical savanna environments: definition, distribution, diversity and development, in *Human Ecology in Savanna Environments* (Ed. D. R. Harris), pp. 3–27, Academic Press, London.

Hopcraft, D. (1975). Productivity comparison between Thomson's gazelle and cattle, and their relation to the ecosystem in Kenya, Ph.D. Thesis, Cornell University.

Hopcraft, D. (no date). Wildlife ranching: a new concept of land use, Nairobi, Kenya (pamphlet).

Houston, D. C. (1979). The adaptations of scavengers, in *Serengeti: Dynamics of an Ecosystem*. (Eds. A. R. E. Sinclair and M. Norton-Griffiths), pp. 263–86, Chicago University Press.

Huygelen, C. (1955). Eland and their possible economic importance (Flem.), *Bull. Agric. Congo Belge.*, **46**, 351.

Jarman, P. J. (1974). The social organisation of antelope in relation to their ecology, *Behaviour*, **48**, 215–67.

Jelliman, A. R. (1913). Game domestication in Rhodesia, *Rhod. Agric. J.*, **10**, 719–20.

Joubert, D. M. (1968). An appraisal of game production in South Africa, *Trop. Sci.*, **10**, 200–11.

Kerr, M. A. and Roth, H. H. (1970). Studies on the agricultural utilisation of semi-domesticated eland (*Taurotragus oryx*) in Rhodesia: 2. Feeding habits and food preferences, *Rhod. J. Agric. Res.*, **8**, 149–55.

Kerr, M. A., Wilson, V. J. and Roth, H. H. (1970). Studies on the agricultural utilisation of semi-domesticated eland (*Taurotraqus oryx*) in Rhodesia, *Rhod. J. Agric. Res.*, **8**, 71–7.

Krebs, J. R. and Coe, M. J. (1985). Sahel famine: an ecological perspective, *Nature (Lond.)*, **317**(6032), 13–14.

Kyle, R. (1972). *Meat Production in Africa—The Case for New Domestic Species*, Veterinary School, University of Bristol, UK.

Kyle, R. (1987). *A Feast in the Wild*, Kudu, Oxford.

Lamb, H. H. (1982). *Climate, History and the Modern World*, Methuen, London.

Lamprey, H. F. (1963). Ecological separation of the large mammal species of the Tarangire Game Reserve, Tanganyika, *E. Afr. Wildl. J.*, **1**, 63–92.

Lamprey, H. F. (1983). Pastoralism yesterday and today: the overgrazing problem, in *Tropical Savannas* (Ed. F. Bourliere), Elsevier Scientific Publications, Amsterdam.

Laws, R. M. (1965a). Aspects of reproduction of the African Elephant: *Loxodonta africana*, *J. Reprod. Fert.*, Suppl. 6, 193–217.

Laws, R. M. (1965b). The Tsavo Research Project, *J. Reprod. Fert.*, Suppl. 6, 495–531

Ledger, H. P. (1968). Body composition as a basis for a comparative study of some East African mammals, *Symp. Zool. Soc. Lond.*, **21**, 289–310.

Leuthold, W. (1977). Spatial organisation and strategy of habitat utilisation of elephants in Tsavo National Park, Kenya, *Z. Saugetierk.*, **42**(6), 358–79.

Leuthold, W. and Sale, J. B. (1973). Movements and patterns of habitat utilisation of elephants in Tsavo National Park, Kenya, *E. Afr. Wildl. J.*, **11**, 369–84.

Lewis, J. G. (1977). Game domestication for animal production in Kenya: activity patterns of eland, buffalo and zebu cattle, *J. Agric. Sci. Camb.*, **89**, 551–63.

Lewis, L. G. (1978). Game domestication for animal production in Kenya: shade behaviour and factors affecting the herding of oryx, buffalo and zebu cattle, *J. Agric. Sci. Camb.*, **90**, 587–95.

McNaughton, S. J. (1979). Grassland–herbivore dynamics, in *Serengeti: Dynamics of an Ecosystem* (Eds. A. R. E. Sinclair and M. Norton-Griffiths), pp. 46–81, Chicago University Press.

McNaughton, S. J. (1985). Ecology of a grazing ecosystem: the Serengeti, *Ecol. Monog.*, **55**(3), 259–94.

Maddock, L. (1980). The 'migration' and grazing succession, in *Serengeti: Dynamics of an Ecosystem* (Eds. A. R. E. Sinclair and M. Norton-Griffiths), pp. 46–81, Chicago University Press.

Malpas, R. C. (1978). The ecology of the African elephant in Rwenzori and Kabalega Falls National Parks, Uganda, Ph.D. Thesis, University of Cambridge, UK.

Martin, P. S. (1973). The discovery of America, *Nature*, **112**, 339–42.

Myers, N. (1972). National parks in savanna Africa, *Science*, **178**, 1255–63.

Myers, N. (1979). *The Sinking Ark*, Pergamon Press, Oxford.

Ogot, B. A. (1981). *Historical Dictionary of Kenya*, Scarecrow Press, New York.

Olang, M. O. (1986). Plant biomass production and large herbivore utilisation in the Kora National Reserve, in *Kora: An Ecological Inventory of the Kora National Reserve, Kenya* (Eds. M. Coe and N. M. Collins), pp. 129–34, Royal Geographical Society, London.

Ormerod, W. E. (1976). Ecological effect of the control of African Trypanosomiasis, *Science*, **191**, 815–21.

Parker, I. and Amin, M. (1983). *Ivory Crisis*, Chatto and Windus—The Hogarth Press, London.

Parker, I. S. C. and Graham, A. D. (1971). The ecological and economic basis for game ranching in Africa, in *The Scientific Management of Animal and Plant Communities for Conservation* (Eds. E. Duffey and A. S. Watt), pp. 393–404, Blackwell Scientific Publishers, Oxford.

Pennycuick, C. J. (1979). Energy costs of locomotion and the concept of 'foraging radius', in *Serengeti: Dynamics of an Ecosystem* (Eds. A. R. E. Sinclair and M. Norton-Griffiths), pp. 164–84, Chicago University Press.

Peters, R. H. (1980). Useful concepts in predictive ecology, *Synthese*, **43**, 257–69.

Phillipson, J. (1975). Rainfall, primary production and 'carrying capacity' of Tsavo National Park (East), Kenya, *E. Afr. Wildl. J.*, **13**, 171–201.

Phillipson, J. (1977). Wildlife—a clue to balancing the environmental budget in Kenya, *Post (Kenya)*, **11**(6), 3–8.

Pratt, D. J. and Gwynne, M. D. (Eds.) (1977). *Rangeland Management and Ecology in East Africa*, Hodder and Stoughton, London.

Pullan, R. A. (1984). The use of wildlife as a resource in the development of Zambia, in *Natural Resources in Tropical Countries* (Ed. Ooi Jin Bee), pp. 267–325, Singapore University Press, Singapore.

Richter, W. von (1969). Report to the Government of Botswana on a survey of the wild animal trade and skin industry, UNDP/FAP No. TA 2637, UN/FAO, Rome.

Rosenzweig, M. L. (1968). Net primary production of terrestrial communities: predictions from climatological data, *Am. Nat.*, **102**, 67–74.

Silberbauer, G. B. (1981). *Hunter and Habitat in the Central Kalahari Desert*, Cambridge University Press, Cambridge, UK.

Sinclair, A. R. E. (1974). The natural regulation of buffalo populations in East Africa: Part 4, The food supply as a regulating factor and competition, *E. Afr. Wildl. J.*, **12**, 291–311.

Sinclair, A. R. E. (1977). *The African Buffalo*, Chicago University Press.

Sinclair, A. R. E. (1979). Dynamics of the Serengeti ecosystem, in *Serengeti: Dynamics of an Ecosystem* (Eds. A. R. E. Sinclair and M. Norton-Griffiths), pp. 1–30, Chicago University Press.

Sinclair, A. R. E. (1983). The adaptations of African ungulates and their effects on community functions, in *Tropical Savannas* (Ed. F. Bourliere), Elsevier, Amsterdam.

Sinclair, A. R. E. and Fryxell, J. M. (1985). The Sahel of Africa: ecology of a disaster, *Can. J. Zool.*, **63**, 987–94.

Skinner, J. D. (1966). An appraisal of the eland (*Taurotragus oryx*) for diversifying and improving animal production in Southern Africa, *Afr. Wildl.*, **20**, 29–40.

Skinner, J. D. (1967). An appraisal of the eland as a farm animal in South Africa, *Anim. Breed. Abstr.*, **35**, 177–86.

Skinner, J. D. (1973). An appraisal of the status of certain antelope for game ranching in South Africa, *Z. Tierzucht. Zucht. Biol.*, **90**, 263–77.
Stanley-Price, M. R. S. (1976). Feeding studies of oryx on the Galana Ranch, *Afr. Wildl. Leader. Found. News*, **11**, 7–11.
Stiles, D. and Muro-Hay, S. C. (1981). Stone cairn burials at Kokurmatakore, Northern Kenya, *Azania*, **16**, 151–66.
Stoddart, D. and Walsh, R. P. D. (1979). Long-term climatic change in the Western Indian Ocean, *Phil. Trans. Roy. Soc. B.*, **286**, 11–23.
Talbot, L. M., Payne, W. J. A., Ledger, H. P., Verdcourt, L. D. and Talbot, M. H. (1965). *The Meat Production of Wild Animals in Africa*, Comm. Agric. Bureaux, Farnham, UK.
Toumalin, C. (1988). Smiling in the Sahel, *New Scientist*, **120**(1638), 69.
Tyrrell, J. G. and Coe, M. J. (1974). The rainfall regime of the Tsavo National Park, Kenya and its potential phenological significance, *J. Biogeog.*, **1**, 187–92.
von la Chevallierie, M. (1970). Meat production from wild ungulates, *Proc. S. Afr. Soc. Anim. Prod.*, **9**, 73–88.
Wade, N. (1974). Sahelian drought: no victory for Western science aid, *Science*, **185**, 234–7.
Wall, R. and Strong, L. (1987). Experimental consequences of treating cattle with the antiparasitic drug Invermectin, *Nature*, **327**(6121), 418–21.
Walter, H. (1954). Le facteur eau dans les régions arides et sa signification pour l'organisation de la végétation dans les contrées sub-tropicals, in *Les Divisions Ecologiques du Monde*, pp. 27–39, Centre Nat. de la Res. Scient., Paris.
Watson, R. M. (1969). A survey of the large mammal population in South Turkana, *Geog. J.*, **135**(4), 529–46.
Watson, R. M. (1972). *Results of Aerial Livestock Surveys of Kaputei Division, Samburu Districts, and North-Eastern Province*, Statistics Division, Min. of Finance and Planning, Republic of Kenya.
Western, D. (1979). Size, life history and ecology in mammals, *Afr. J. Ecol.*, **17**, 185–204.
Western, D. (1980). The ecology of past and present mammal communities, in *Fossils in the Making* (Ed. A. K. Behrensmeyer and A. P. Hill), pp. 41–54, Chicago University Press, Chicago.
Whittaker, R. H. (1970). *Communities and Ecosystems*, Collier–Macmillan, London.
Winstanley, D. (1973). Rainfall patterns and general atmospheric circulation, *Nature (Lond.)*, **245**, 190–4.

CHAPTER 9

Conservation and Management of Water Resources

ASIT K. BISWAS

INTRODUCTION

Water is an essential requirement for desert reclamation, since the bio-productivity of desert-like areas cannot be enhanced and sustained without a regular and reliable water supply. The very fact that an area is a desert indicates a shortage of water for effective bioproduction.

Desert or desert-like conditions exist in all continents, though the extent of such land mass varies significantly from one continent to another. If a continent like Africa is considered, on the basis of climatic analyses, 29% of its land area is under desert-like conditions (annual rainfall less than 100 mm), 17% is arid (100–400 mm/yr), 8% semi-arid (400–800 mm/yr), 10% dry sub-humid (800–1200 mm/yr), 20% moist sub-humid (1200–1500 mm/yr) and the balance of 16% of land area is humid (annual rainfall over 1500 mm) (Biswas, 1987). Generally speaking, the dry sub-humid zone of Africa (growing period 120–179 days) has adequate moisture availability for rainfed production of staple food crops like millet, sorghum and maize. Overall, considering the total African land area, approximately 25% of it can be considered to be suitable for rainfed production and another 10% is marginally suitable. The remaining 65% of the land is unsuitable.

Under such different climatic conditions, the availability of surface water varies tremendously from one part to another. In the region of Sahara and the Horn of Africa, there is no surface water since there is no runoff. In the Sudano-Sahelian region, extending from Senegal to Somalia, the average runoff is up to 10% of rainfall, and it increases to more than 20% in the wet tropical highlands of Ethiopia. In contrast, the Congo River Basin, which at 4 million km^2 covers nearly 16% of sub-Saharan Africa, has a mean annual

Techniques for Desert Reclamation
Edited by A. S. Goudie
© 1990 John Wiley & Sons Ltd.

discharge of 1325 km³, which accounts for 55% of the mean annual discharge of that region.

Such variations in water availability, both in terms of space and time, mean that water management and conservation must play a very important role in arid and semi-arid areas if the land is to have the capacity to support the existing human and livestock population as well as the expected increments in the future. Thus, in a country like Kenya, it is estimated that 750 mm or more of rainfall is necessary for crop production, but only 15% of the country receives such rainfall. With one of the highest population growth rates in the world, Kenya thus must practise sound land and water management policies so that its arid and semi-arid land areas can sustain their bioproductivity. Without such policies, there is a real danger of increasing desertification, as has been noted in many areas of Africa and other parts of the world. As Tolba (1987) has noted, during the last 100 years, a 150-km wide belt of productive land on the southern edge of the Sahara has turned completely unproductive. Since 1968, one quarter of Africa's semi-arid pasturelands—the main source of meat—has also been rendered unproductive through accelerated desertification.

WATER MANAGEMENT

Even though good water management is critical for arid and semi-arid areas, the level of current management practices leaves much to be desired. It is imperative that whatever water resources are available in arid and semi-arid areas, these must be used efficiently.

To develop efficient water management plans for any area, an important requirement is the availability of reliable assessment of the resource. Since distribution of water varies with time and space, reliable forecasts of water availability can only be made on the basis of adequate data over a reasonable period of time. Based on such data, long-term water management plans can be established. However, in many arid and semi-arid developing countries, such data are not available, or if available, they are often only for a limited period of time. Unfortunately, the present economic situation in Africa and Latin America has meant that the hydrologic data collection systems in many countries have actually undergone deterioration instead of improvement. Data reliability still continues to be a major problem.

If a reasonable range of data is available, several alternatives can be explored in terms of efficient water management. These alternatives are not necessarily mutually exclusive. Among these are the following:

(a) efficient use of water available from the projects that have already been constructed;

(b) use of large-scale irrigation development, including long-distance water transfer;

(c) development of small-scale systems like check dams and tanks; and
(d) reuse of wastewater.

Since much literature exists at present on efficient water use and large-scale water development, these two alternatives will be discussed only briefly.

EFFICIENT WATER USE

Studies carried out for the United Nations Water Conference held in Mar del Plata, Argentina, in 1977 (Biswas, 1978), indicated that globally agriculture accounted for an average of 80% of all water use, and that the agricultural sector is an inefficient user of water. On a global basis it was estimated that 1.3×10^{12} m^3 of water was used for irrigating crops, for which 3×10^{12} m^3 of water had to be withdrawn. This indicates that 57% of water withdrawn was lost in the distribution system. Since this estimate was based on official government statistics, it is highly likely that the real situation is much worse.

Even if the estimate was reasonably reliable, it should not be assumed that 43% of the water reaching irrigated fields was efficiently used. Overirrigation is endemic in all arid and semi-arid countries, though the actual efficiency of use varies significantly between countries as well as projects within a country. This means water, which could have been used for other productive purposes, is actually being wasted. In addition, excessive water use often has contributed to development of adverse environmental problems like waterlogging and salinity, which results in the reduction of the bioproductivity of irrigated land. The problems of waterlogging and salinity are discussed in detail in Chapter 4.

It should be noted that, as a rule, it is more economic to obtain additional water by improving the efficiency of water use from existing water projects than from building new ones. Also, the time required to plan and to build new schemes is significantly longer: the efficiency of existing projects can be improved more quickly. However, a word of caution is in order. Water-use patterns in many areas have evolved over decades of practice, and all the indications are that changing water-use patterns may be technically simple but in reality may prove to be a very difficult and complex process due to socioeconomic and political constraints.

LARGE-SCALE IRRIGATION DEVELOPMENT, INCLUDING LONG-DISTANCE WATER TRANSFER

If rivers and suitable dam sites exist, it is possible to develop an extensive irrigation system which could reclaim desert areas. For example, construction of the Aswan Dam enabled Egypt to reclaim desert areas for agricultural production. Similar examples are available in many other arid countries.

Another alternative available for desert reclamation has been to transfer water over long distances by a canal, thus transferring water from a water-surplus to a water-deficient area. A good example of such a long-distance water-transfer project for desert reclamation is the Indira Gandhi Canal Project (formerly known as the Rajasthan Canal Project) in India. The project transfers water from the state of Punjab to reclaim part of the Rajasthan Desert.

The first phase of the project has now been successfully completed. This phase covers an area of 200 000 ha, where irrigation was gradually introduced over the period of 1961 to 1978. In the areas where irrigation started in the early 1960s, the desert land has now been completely transformed into an agricultural area. Introduction of water, along with programmes on dune stabilization, pasture development and afforestation, have made the first phase an effective desert reclamation project.

The second phase of the project, covering another 246 000 ha, appears to be running into some problems, especially in terms of adequate water availability, cost per hectare of reclamation and the difficulty of attracting settlers to a very inhospitable terrain and climate.

It should be noted that long-distance water transfer is a costly alternative for desert reclamation, and can be economically and environmentally justified for a very few project areas.

SMALL-SCALE SYSTEMS: CHECK DAMS AND TANKS

Small-scale decentralized systems can be used effectively to increase the bioproductivity of the arid and semi-arid areas. Two most important alternatives are check dams and tanks.

Check Dams

Check dams are generally small and low dams which are built across gullies and streams to store flood runoff. The practice has successfully been used for several centuries in rural areas of several countries like India, China, Sri Lanka, the United States and Mexico. Basically check dams store flood water upstream, which can be used subsequently for irrigation and live-stock. Depending on their locations and availability of alternative sources, water stored behind check dams may also be used for human consumption, especially for bathing and washing.

Check dams could range from relatively simple structures built with stones, gravels and clay to fairly elaborate and sophisticated rockfill dams with concrete spillways. Many of the early check dams were simple structures that were built across narrow valleys, having somewhat impervious rock or soil strata. These dams required very minor changes in the local topography, and accordingly could be constructed relatively quickly with low financial invest-ment as well as limited expertise and labour input. These dams not only control the flow of water but also silt that is carried by flood waters. With the reduction in flow velocities by the presence of these dams, the rates of soil erosion are also reduced. Construction of a series of check dams on a gully or stream can significantly reduce the overall rate of soil erosion. Furthermore, as the flow velocities are reduced, silt present in flood water is deposited behind such dams.

As the silt deposited on the river bed increases every year, after a period of time a very fertile area is available for cultivation, especially when the stored water disappears. Thus, check dams are structures that can not only harvest seasonal flood waters but also contribute to soil and moisture conservation.

Large check dams were constructed in Mexico during the eighteenth and the nineteenth centuries. These were generally medium-size masonry struc-tures, with broad bases and supporting buttresses, which provided water for the haciendas. Lateral spillways were often used, and water stored was adequate for cultivating approximately 40 ha of land.

At present some very sophisticated check dams are being constucted in the many arid desert-like areas. One such important dam now under construction is on the Wadi Karm, Sinai, Egypt. This sophisticated rockfill dam, with a concrete spillway, will store the rare flash floods of the wadi, which occur very irregularly. Under such very arid conditions, the reservoir is likely to be dry for a significant part of the year. Thus, water stored can be used for irrigation for one crop season per year.

While check dams have been used for centuries, their use in recent years for water and erosion control is receiving increasing attention in countries as diverse as China, Nepal, India and Ethiopia. In China, check dams have been very successfully used to increase the agricultural production potential in many areas. Similarly, in Nepal, under the Small Farmers Development Programme, check dams are now being extensively used for water storage for small-scale irrigation.

A good example of the use of check dams for water and erosion control can be found in the Juiyuan Gully in Suide County, Shanxi Province, China. The Gully is a small tributary of the Wuding River, which in turn is a tributary of the Huang He (Yellow River). The length of the main channel of the Gully is 18 km and it has a catchment area of 70.1 km^2, of which 2130 ha is used for agricultural production. Some 10 000 people live in this rural area and agriculture is the main source of livelihood.

Water and erosion control was the most serious problem facing the Jiuyuangou Peoples' Commune in the Gully catchment area. The catchment has a high gully density of 5.34 km/km^2. Before a control programme was initiated, the rate of annual soil erosion from the catchment was estimated at 1.27 million metric tonnes, which was equivalent to an average soil loss of

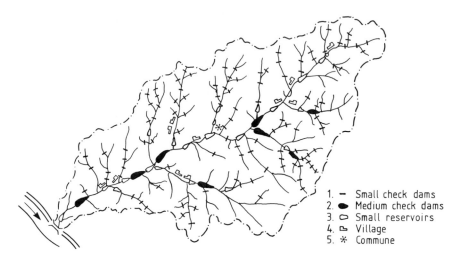

Figure 9.1   Use of check dams in the Jiuyuan Gully Catchment Area, China

18 116 metric tonnes/km². Because of the high silt content of the water in the gully catchment, it could not be used efficiently for irrigation.

The Commune initiated a combined programme of contour farming and check dams. By 1974, 727 ha of land had been provided with contour farming. In addition, 311 small and medium check dams were constructed, which primarily acted as silt traps, and 30 small reservoirs were built to store a total of 1.18 million m³ of water for irrigation. The general plan of the Jiuyuan Gully catchment is shown in Figure 9.1.

The construction of this series of small- and medium-size check dams had a remarkable impact on the water use and erosion rates of the area. The irrigated area increased to 170 ha, which was eleven times the preconstruction period figure. The flood peak in the Gully was 90% less than before and the average annual soil loss decreased by 770 100 metric tonnes, which was a reduction of nearly 60%. This meant that the total agricultural production in the area increased by 2.3 times within a period of only two decades.

Because check dams are small and widely dispersed over rural areas in many arid countries, it is not possible to comment on their overall efficiency. Furthermore, the check dams are often constructed by local people, based on past experience and broad rules of thumb. Thus, there are numerous types of such dams, based on different 'design' parameters, located in an immense variety of site-specific topographical and other physical conditions. Their maintenance often differs from one location to another. In addition, not even a single country has made a national survey of these dams. Under these conditions, only some general comments can be made on their advantages and limitations.

Check dams, when they are properly designed, constructed and maintained, can be considered to be a very useful small-scale alternative for water control in rural areas of arid and semi-arid countries. They are easy to design, construct and maintain. Labour and capital requirements are minimal, certainly significantly less than other sophisticated hydraulic structures. This means individual households or small communities can afford to build these dams without external assistance. Foreign exchange is generally not necessary, which could be an important consideration for many debt-laden developing countries at present. These simple structures can be constructed within a very short period of time, compared to large dams where the gestation periods are often more than a decade. Large-scale centralized institutions are not necessary for their construction, operation and maintenance. Also a series of such dams in an area can be developed incrementally.

Check dams have many limitations as well. They provide unreliable and discontinuous supplies of water, which means communities must have access to other alternative sources of water. Because of their decentralized nature, they often suffer from poor quality of design, improper construction and inadequate maintenance. Frequent repairs are necessary, but these repairs can be carried out quickly and within a limited cost. Many check dams are very vulnerable in terms of water-quality contamination, and they often act as the main foci of water-borne diseases, especially those transmitted by mosquitoes.

Tanks

Tanks are basically reservoirs created behind small earth dams built across streams to collect rainwater. In India and Sri Lanka, these reservoirs are called *tanks*, and have been used for harvesting rainwater for centuries. The surface area covered by these tanks could vary enormously—from a few hectares to thousands of hectares. The water from the tanks is used for all purposes: domestic requirements as well as for irrigation and livestock. Many tanks are also used for aquaculture.

Tanks normally contain a weir, over which excess water is discharged downstream. The height of the weir thus determines the water-storage level of the tank. The weir can be located along the dam itself or at an appropriate location somewhere around the tank perimeter. While tanks can generally store flows due to normal rainfall, excessive rainfall could overtop the earthen dams. This could result in serious erosion which needs to be repaired after the flood season.

Another type of small tank can also be seen in many parts of India. These are different in the sense that they are created not by earthen dams but by excavation in a flat area. These small tanks are not along a river, but generally at some distance from it. They are very popular in the rural areas of provinces

Table 9.1   Irrigation in India by tanks, 1963

Province	Total number of tanks	Area irrigated by tanks (10^3 ha)	Contribution of tanks to irrigation (%)	Tank density per km^2
Andhra Pradesh	56 700	1232	41.5	0.2
Assam	0	0	0	0
Bihar	27 800	276	14.0	0.16
Maharashtra	48 100	191	17.7	0.16
Gujarat	20 400	16	2.2	0.10
Jammu and Kashmir	0	0	0	0
Kerala	1 500	45	13.4	0.04
Madhya Pradesh	40 200	151	15.9	0.09
Karnataka	36 500	355	40.2	0.19
Orissa	1 700	412	40.1	0.01
Tamil Nadu	31 400	928	37.7	0.24
Punjab	200	6	0.2	0.004
Rajasthan	N/A	165	9.2	N/A
Uttar Pradesh	140 000	413	8.2	0.48
West Bengal	100 000	364	27.2	1.14
Total	504 500	4554		

Source: Planning Commission, 1966.

like West Bengal or Orissa in India, where individual households often have such tanks for watering small plots of land, and for domestic use and aquaculture. However, with the value of land rising rapidly, many such small tanks are now being filled up with soil so that alternate and more economic use could be made of the land.

The Indian Planning Commission (1966) made an estimate in 1963 of tanks that were being used in different parts of the country. This is shown in Table 9.1. It was estimated that the total number of tanks in the country exceeded 500 000 and altogether they provided irrigation water for some 4.55 million ha of land. The average area irrigated by a tank was 9 ha and the average water-holding capacity of a tank was 351 000 m^3. Storage capacities naturally depend on a variety of conditions, among which are topographical conditions, water requirements and expected runoff. Some of the tanks, especially in the southern province of Tamil Nadu, are quite large and can store as much as 100 million m^3 of water. In the three southern Indian states of Andhra Pradesh, Tamil Nadu and Karnataka, nearly 9.7% of the annual rainfall is stored in the tanks (UNEP, 1983). Thus, tanks could play an important role in providing irrigation water to increase agricultural production.

Tanks have not received enough research attention so far, but based on specific individual studies some comments can be made on the efficiency of water storage and use by tanks. They need continual maintenance if they are to function properly. Overtopping and subsequent breaching of the earthen perimeter wall is a continuing threat, and generally the weirs provided cannot handle high flows. The report prepared by the Indian Planning Commission (1966) estimated that on an average the weirs can discharge only about a quarter of the expected maximum discharge. The Commission also noted the failure of tanks due to excessive seepage and resulting piping. In those rivers where a successive series of tanks have been constructed, failure of one tank due to overtopping or any other reason could seriously endanger the safety of the ones downstream.

The losses due to evapotranspiration and seepage could be quite substantial. Evaporation losses are often high, because, in contrast to reservoirs, tanks are rather shallow and thus have a large surface area in comparison to the volume of water stored. This high surface area increases evaporation losses. Presence of aquatic weeds in many tanks further increases water losses through transpiration. Evaporation losses in arid areas could easily reach as high as 3 m every year.

Observed data on seepage losses from tanks are not available, though it has been estimated to vary from 1 to 4 m of water per year. Naturally the seepage losses will depend primarily on the soil characteristics. As a general rule, seepage losses are high during the early life of a tank but then reduce with time. This is because fine silts tend to act as a sealant, thus making the soil more impervious and reducing the resulting water loss due to seepage.

Water-quality deterioration and silting are two major problems faced by tanks. Since water is stagnant for much of the year, and tanks are used for all purposes like bathing, washing and livestock washing, the water quality tends to deteriorate until the onset of the rainy season, when the extra water has a diluting effect. Floodwater increases the turbidity of the tank temporarily, and on a longer-term basis contributes to a loss of storage due to silting. Soil erosion in arid countries is often high. When the rains start after a hot summer season, the vegetative cover is usually at a minimum, which means high rates of soil erosion. Part of this eroded soil becomes deposited in the tanks. However, unlike reservoirs, many tanks become dry during the summer season. Also all the water can be drained on purpose. These possibilites allow for the clearing of the accumulated silt. This, however, may not happen since the tanks are often communal property, whereas silt clearance requires individual efforts of the beneficiaries. People are generally reluctant to contribute their labour or funds for silt clearance as the benefit accrues to the whole community and not exclusively to the individuals associated with the cleaning process. Their communal nature is also an important reason as to why tank maintenance could leave much to be desired. Accordingly, unless there is strong communal leadership in the area, overall tank management tends to become inadequate.

Tanks, like other small-scale irrigation works, have some special attractions. They do not require major investments in physical infrastructures and generally have no foreign exchange requirements. These schemes can be developed at relatively low costs and thus can be cost-effective for a wide variety of crops, including some basic staples (Biswas, 1986). Since these small-scale projects are less complex and relatively simple to construct, less time is required for planning and construction and hence they can contribute to agricultural production quite quickly. Farm-level investments are also low, which means smallholders can afford this form of water control.

Cost per hectare of irrigation development thus is reasonable. Naturally the cost could vary from one project to another, depending on site-specific conditions like complexity of schemes, terrain conditions, etc. The average cost in India, however, has been estimated at around $1000 per ha. It should, however, be noted that rigorous cost analyses for tanks are simply not available. In addition, the beneficiaries may provide some labour and/or resources for the construction of the projects and also may play a useful role in their operation and maintenance. This type of beneficiary costs generally does not appear to be considered in cost calculations.

Tanks have many advantages, but they also suffer from diseconomics of scale, poor efficiency and quality control, and lack of governmental interest and supervision due to their decentralized nature, small size and predominantly rural locations.

REUSE OF WASTEWATER

Reuse of wastewater for desert reclamation and enhancing agricultural production in desert-like areas has received increasing attention in recent years, especially during the past decade. Wastewater is now being increasingly used to enhance bioproductivity of very arid areas in countries as diverse as China, India, Egypt, Jordan, Saudi Arabia, Kuwait, Oman and Cyprus. To a major extent the increasing interest in wastewater reuse has been a direct byproduct of the International Water Supply and Sanitation Decade, which was proclaimed by the United Nations for the period 1981–90. With the construction of centralized sewer systems and water-treatment works for urban centres, many arid countries suddenly realized that this was a 'new' source of water which could be used for productive purposes.

Currently all arid countries of the Middle East and North Africa are embarking upon ambitious programmes on wastewater treatment and reuse. For these countries with large desert areas, water is already a serious constraint for further development. Conventional sources of water have already been committed or are about to be fully committed, and no additional sources of water generally exist for further agricultural development.

Generally speaking, for most arid and semi-arid countries, reuse of wastewater may have a greater impact on future water availability than any other technological solutions for increasing water supply such as water harvesting, weather modification or desalination. Treated wastewater can be used for irrigation, industrial purposes and groundwater recharge (Biswas and Arar, 1988a, 1988b). Furthermore, as various agricultural and industrial demands are met by wastewater, more fresh water could be made available for domestic purposes.

As environmental requirements become more stringent in many arid countries, primary, secondary and tertiary wastewater treatment plants are becoming more and more common. In some countries like Saudi Arabia, wastewater is being ozonated prior to any reuse, which from a health viewpoint is somewhat unnecessary. This means that treated wastewater has become of fairly good quality for use for many purposes. If proper groundwater recharge is practised, treated wastewater can supplement nearly all the various uses of groundwater, including being a potential source of municipal water supply.

Two important factors should be noted in terms of wastewater reuse. First, in terms of total quantity, wastewater available for use is somewhat limited. Second, in arid countries, direct use of wastewater can generally be best made near centres of population, where it is produced. This will ensure that transmission losses due to seepage and evaporation would be minimal, and also better quality control can be exerted on its use. In Aqaba, Jordan, treated wastewater produced is now being used in nearby areas for desert reclamation. Similarly, in many Middle Eastern and North African countries, treated wastewater is being used to develop green areas within and around cities.

There are some significant advantages in using treated wastewater, among which are the following:

(a) The marginal cost of providing additional good-quality water of the same volume as the wastewater generated will generally be higher than using that wastewater.

(b) Since sewage treatment works are necessary because of health and environmental requirements, treated wastewater would be produced irrespective of whether or not it is used. Thus, it makes economic sense to use this additional resource as beneficially and efficiently as possible.

(c) Wastewater can be especially beneficial for desert reclamation, where the soil often lacks humus and nutrients. Since treated wastewater contains high concentrations of nutrients, especially nitrogen and phosphorus, its use under desert-like conditions enhances bioproductivity by providing both water and fertilizer.

(d) Salt-water intrusion in the coastal arid areas can be prevented by recharge of groundwater with treated wastewater.

(e) Volume of wastewater produced does not fluctuate widely like streamflow. Accordingly it is a more reliable and stable source of water.

Wastewater reuse also has some special health and environmental implications. Among the issues that need to be considered, the following are worth noting:

(a) *Aesthetic:* Three aesthetic conditions should be noted: colour, odour and foam. The colour of effluents indicates their state of degradation and the presence of algae. The colour is usually black during septic conditions and green due to the presence of algae. The presence of an undue amount of algae may create problems for drip irrigation. Odour problems due to anaerobic processes taking place, could be a concern if they occur near residential areas. Foaming is due to the type of detergents used, and is primarily considered to be an environmental nuisance.

(b) *Solids:* The extent of organic and inorganic solids, both in dissolved and suspended forms, is an important consideration for irrigation purposes. Particle-size distribution of suspended solids may preclude certain forms of irrigation, like sprinkler and drip, and may also affect the permeability of the soil if long-term groundwater recharge is to be considered.

(c) *Nutrients:* Treated wastewater contains a significant amount of nutrients in the forms of nitrogen, phosphorus and potassium, which can foster plant growth and reduce the extent of commercial fertilizers that need to be applied to the soil. The concentrations of the various nutrients are affected by the type and extent of treatment used.

(d) *Salinity:* If the salinity of wastewater is high and the water is used for irrigation, salts tend to concentrate in the soil, and the water is often used up by evapotranspiration. As salts begin to accumulate in the root zones, the yields of crops grown begin to decline.

(e) *Bacteria:* The bacterial quality of wastewater is an important consideration for irrigation and groundwater recharge. Types of crops that can be irrigated depend on the bacterial quality of the wastewater. Similarly, some restrictions may be imposed in terms of termination of irrigation before harvesting.

(f) *Toxicity:* Plants may accumulate certain constituents from the irrigation water, which may create toxicity within them. The toxic substances that need to be monitored are sodium, chloride and boron.

(g) *Metals:* The presence of certain trace metals in wastewater may constitute a hazard to soil organisms, plants, people and animals consuming the various plant products.

While interest in reusing wastewater for irrigation is undeniable in most arid and semi-arid countries, the attempts made so far have generally been somewhat haphazard. Many countries have often felt that since limited water availability is a constraint to development, any additional water that can be harnessed is a positive step. Good intentions alone, however, are not enough: they must be supported by proper planning and management of waste-

Table 9.2   Framework for wastewater reuse and/or groundwater recharge in arid areas

---

1. Nature of the problem
   (a) How much wastewater will be produced and what will be the seasonal distribution?
   (b) At what places will wastewater be produced?
   (c) What will be the characteristics of wastewater that will be produced?
   (d) What are feasible alternative disposal possibilities?
2. Legal feasibility
   (a) What uses of wastewater are possible under national and/or state regulations, if they exist?
   (b) If no regulations exist, what uses seem feasible under WHO and FAO guidelines for irrigation?
   (c) What are the prevailing water rights and how will these be affected by wastewater use?
3. Technical feasibility
   (a) Is the quality of treated wastewater produced acceptable for restricted or unrestricted irrigation?
   (b) How much land is available or required for wastewater irrigation?
   (c) What are the soil characteristics of land to be irrigated?
   (d) What are the present land-use practices? Can these be changed?
   (e) What types of crops can be grown?
   (f) How do crop-water requirements match with seasonal availability of wastewater?
   (g) What types of irrigation techniques can be used?
   (h) If groundwater recharge is a consideration, are the hydrogeological characteristics of the study area suitable?
   (i) What will be the impact of such recharge on groundwater quality?
   (j) Are there additional health and environmental hazards that should be considered?
4. Political and social feasibility
   (a) What have been the political reactions to past health and environmental hazards that may have been associated with wastewater reuse?
   (b) What is the public perception of wastewater reuse?
   (c) What are the attitudes of influential people in areas where wastewater will be reused?
   (d) What are the potential benefits of reuse to the community?
   (e) What are the potential risks?
5. Economic feasibility
   (a) What are the capital costs?
   (b) What are the operation and maintenance costs?
   (c) What is the economic rate of return?
   (d) What are the costs of development of effluent-irrigated agriculture, e.g. cost of conveyance of wastewater to the irrigation site, land-levelling, installation of irrigation system, agricultural inputs, etc.?
   (e) What are the benefits from the effluent-irrigated agricultural system?
   (f) What is the benefit–cost ratio for the irrigation project?
6. Manpower feasibility
   (a) Is adequate local manpower available for adequate operation and maintenance of:
       wastewater treatment,
       irrigation and groundwater recharge works,
       agricultural facilities,
       health and environmental control aspects?
   (b) If not, what types of training programmes should be instituted?

Table 9.3 Intercomparison of wastewater reuse for irrigation, groundwater recharge and overland flow

Factors	Irrigation	Groundwater recharge	Overland flow
Treatment	Primary to secondary	Untreated to primary	Untreated to primary
Consistently good operation of treatment plants	Critical	Not critical	Not critical
Water quality	High	Medium to low	Medium to low
Land area acquired	High	Low	Medium
Land slope	Up to 6% for surface irrigation; up to 30% for sprinkler and drip irrigation	Not important, but difficult on steep slopes	1–12%
Soil permeability	Moderate	Rapid to very rapid	Low
Soil quality	Medium to good	Not important	Not important
Utilization of water and nutrients	High	None	Medium to low
Monitoring requirements	Extensive	Limited	Limited

water reuse, which will ensure that an optimal cost-effective process can be designed and maintained for each of the site-specific situations under consideration, consistent with health, environmental and institutional constraints. *Ad hoc* planning of projects, without clear ideas about their long-term sustainability, are unlikely to be successful on a long-term basis.

Planning of wastewater reuse projects needs a systematic approach. A systems framework to such a planning process is shown in Table 9.2. The framework suggested is in six phases: nature of the problem, legal feasibility, technical feasibility, political and social feasibility, economic feasibility and manpower feasibility. In each phase, many specific questions need to be asked and answered. In Table 9.2, many of the important questions have been identified. The framework should be considered as a general approach to planning, and the various phases and questions indicated should not necessarily be considered to be sequential, since some of the steps could be simultaneous, and orders may vary, depending on site-specific conditions.

Land application of wastewater has undergone some radical rethinking in recent years. Earlier, the capability of soil to treat wastewater was not adequately recognized and, hence, it was significantly underestimated. Accordingly, often high discharge standards were imposed on the effluents, prior to land application, which unnecessarily increased total treatment costs, overall energy requirements and expertise required. As land application is increasingly being considered as an important and essential component of wastewater treatment and reuse processes in arid and semi-arid countries, the general philosophy has undergone a major change. Now the emphasis has shifted to the application of water-quality standards following land treatment, and if the effluent is to be reused for irrigation immediately following the treatment, the emphasis is on how to ensure that the products to be consumed and the agricultural workers do not face undue health risks.

Table 9.3 shows the various requirements of land application of wastewater in terms of irrigation, groundwater recharge and overland flow, all of which can be practised under arid and semi-arid conditions.

CONCLUDING REMARKS

Proper conservation and management of water resources is absolutely essential for desert reclamation as well as to ensure that productive lands do not undergo desertification. There are several alternatives to improve water management and conservation practices, and the alternative that can work best in one place may not be optimal under different conditions in another location. Management alternatives should therefore be chosen for project-specific conditions.

REFERENCES

Biswas, Asit K. (1978). *United Nations Water Conference: Summary and Main Documents*, Pergamon Press, Oxford.

Biswas, Asit K. (1986). Irrigation in Africa, *Land Use Policy*, **4**(3), 269–85.

Biswas, Asit K. (1987). Water for food production in sub-Saharan Africa, *Geo-Journal*, **15**(3), 233–42.

Biswas, Asit K and Arar, A. (1988a). *Treatment and Reuse of Wastewater*, Butterworths, London.

Biswas, Asit K. and Arar, A. (1988b). Use of marginal water quality water for plant production in Europe, *International Journal of Water Resources Development*, **4**(2), 127–41.

Planning Commission, Committee on Plan Projects, Irrigation Team (1966). *All India Review of Minor Irrigation Works Based on Statewise Field Studies*, Planning Commission, New Delhi, India.

Tolba, M. K. (1987). *Sustainable Development: Constraints and Opportunities*, p. 208, Butterworths, London.

United Nations Environment Programme (1983). *Rain and Stormwater Harvesting in Rural Areas*, Cassell Tycooly, London.

Index